Mariculture In India: Understanding Ecosystems for Sustainable Marine Food Production

Edited by

G. Chelladurai
G.V.N. College (Autonomous), Kovilpatti
Thoothukudi district
Tamil Nadu, India

M. Paripooranaselvi
Department of Zoology
St. Mary's College (Autonomous), Thoothukudi
Tamil Nadu, India

&

B. Makesh Kumar
G.V.N. College (Autonomous), Kovilpatti
Thoothukudi district
Tamil Nadu, India

Mariculture In India: Understanding Ecosystems for Sustainable Marine Food Production

Editors: G. Chelladurai, M. Paripooranaselvi and B. Makesh Kumar

ISBN (Online): 979-8-89881-657-5

ISBN (Print): 979-8-89881-658-2

ISBN (Paperback): 979-8-89881-659-9

Published by Bentham Science Publishers Pte. Ltd. Singapore, in collaboration with Eureka Conferences, USA.

First published in 2026.

Bentham Science Publishers Pte. Ltd.
No. 9 Raffles Place
Office No. 26-01
Singapore 048619
Singapore
Email: subscriptions@benthamscience.net

BENTHAM SCIENCE

CONTENTS

FOREWORD-I

The book "Mariculture in India: Understanding Ecosystems for Sustainable Marine Food Production" emphasizes the vital role of mariculture in global economic and community development. From the coast to the deep ocean floor, marine ecosystems represent the largest habitat system on Earth. This book provides a national forum for key information, ideas, and discussion on all aspects of mariculture.

Through a rapid, thorough, and open-access publishing approach, it presents advanced research that helps readers understand marine life, interactions among organisms, ecosystem functioning, and human interactions with the oceans. The book focuses on experimental work at all levels of biological organization, including ecological modeling and ecosystem studies, both in the laboratory and in the field.

The author has organized the book into diverse chapters, including Introduction of Global Mariculture Production and Development, Present Status, and Prospects of Mariculture, providing readers with insights into the current status, challenges, and future potential of the sector. Further chapters, such as Water Resources for Sustainable Mariculture and Mariculture Pollution and Biodiversity, expand knowledge for researchers and offer guidance for addressing ecological challenges. Chapters on Management of Sustainable Mariculture and Development of Aqua Feed for Mariculture Species present key issues in aquaculture development and offer frameworks for improving sustainability from both socioeconomic and ecological perspectives.

The final chapters, Role of Biotechnology in Sustainable Mariculture and Planning and Regulation of Sustainable Mariculture, explore how biotechnology, through nutrition, feeding stimuli, chemical signals, and the reduction of in-pond chemical oxygen demand, can support sustainable practices. All articles, including original research and reviews, have undergone peer review by subject-matter experts and esteemed editors, ensuring the highest quality for the scientific community.

It is a privilege to write the foreword for this book, which thoroughly examines the structure and development of mariculture in India. The book's wide range of topics, authored by experts in their respective fields, provides a comprehensive overview of ecosystem-based, environmentally friendly mariculture practices. I firmly believe the knowledge presented here will be highly beneficial, thought-provoking, and inspirational for the entire scientific community.

S.B. Sharma
Vice-Chancellor
Millennium University
Blantyre
Republic of Malawi

FOREWORD-II

The rapid development of animal and plant biotechnology and the numerous researches conducted under the auspices of various organizations and private industries, both in this country and abroad, make it challenging to interpret the vast amount of information that accumulates daily.

Mariculture is the cultivation of aquatic animals and plants in marine and estuarine waters. The growth of mariculture provides many social and economic opportunities as well as challenges. Improved food security, higher incomes, and increased job opportunities are some potential benefits for local communities. This book is an effort to consolidate and present the understanding of ecosystems for sustainable marine food. It is divided into eight valuable chapters covering the introduction, global mariculture production, present status, prospects of mariculture development, water resources for sustainable mariculture, mariculture pollution and biodiversity, management of sustainable mariculture, development of aquafeed for mariculture species, the role of biotechnology in sustainable mariculture, and planning and regulation of sustainable mariculture. This work aims to provide students with comprehensive information on all aspects of mariculture to support their academic careers.

As part of the commitment to promoting a healthy environment, faculty from various colleges have compiled and collated ideas from academicians and researchers on mariculture.

I wish the editors and all authors the best of luck.

R. Srinivasan
Member Secretary
Tamil Nadu State Council for Science and Technology DOTE Campus
Chennai, 600025, India

PREFACE

The oceans have long been the cradle of life, sustaining civilizations through their vast biodiversity and invaluable resources. In an era marked by rapid population growth and increasing pressure on natural resources, the need for sustainable and responsible food production from marine ecosystems has become more critical than ever. Mariculture, as an emerging scientific discipline and expanding industry, offers a promising pathway to address food security challenges while preserving the delicate balance of marine environments.

This book, ***Mariculture in India: Understanding Ecosystems for Sustainable Marine Food Production***, has been conceived to provide a comprehensive insight into the ecological, biological, and technological foundations of mariculture in the Indian context. It presents an integrated perspective on the rich marine biodiversity of India, advances in culture technologies, ecosystem-based management approaches, and the socio-economic dimensions of sustainable marine food production. The central emphasis of this work is to bridge scientific knowledge with sustainable practices, ensuring that mariculture contributes not only to nutrition and livelihoods but also to the conservation and responsible stewardship of marine ecosystems. The chapters in this book are contributed by experts who bring together their research findings and practical experiences. Collectively, they offer a valuable resource for students, researchers, academicians, policymakers, and practitioners seeking a deeper understanding of mariculture and its pivotal role in building a sustainable future.

We hope that this volume will inspire continued research, innovation, and responsible practices in marine food production systems, while reinforcing our shared responsibility to protect and sustain the ecosystems that support life. It is our firm belief that promoting sustainable mariculture in India will play a significant role in strengthening national food security, supporting economic development, and conserving marine biodiversity.

We express our sincere gratitude to all contributors, reviewers, and supporting institutions whose cooperation made this work possible. We also thank our readers for their interest and engagement with this important field of study.

G. Chelladurai
G.V.N. College (Autonomous), Kovilpatti
Thoothukudi district
Tamil Nadu, India

M. Paripooranaselvi
Department of Zoology
St. Mary's College (Autonomous), Thoothukudi
Tamil Nadu, India

&

B. Makesh Kumar
G.V.N. College (Autonomous), Kovilpatti
Thoothukudi district
Tamil Nadu, India

List of Contributors

Anathakumar Saminathan	Department of Microbiology, Srimad Andavan Arts and Science College (Autonomous), Trichy 620005, Tamil Nadu, India
Anisha Valsalam	Department of Aquatic Environment Management, ICAR – Central Institute of Fisheries Education, Mumbai, Maharashtra, India
Clara Jeya Geetha J.	Department of Chemistry, Pope's College (Autonomous) (Affiliated to Manonmaniam Sundaranar University), Sawyerpuram 628251, Tamil Nadu, India
Dhivya Antony	Department of Chemistry, Dhanalakshmi Srinivasan Arts and Science (Co-Ed) College (Affiliated to University of Madras), ECR, Mamallapuram, Chennai 603104, Tamil Nadu, India
C. Stella Packiam	Department of Chemistry, A.P.C.Mahalaxmi College for Women, Thoothukudi, Tamil Nadu, India
Dhandayuthabani Rajendiran	Department of Chemistry & Biochemistry, C. Abdul Hakeem College (Autonomous) Melvisharam 632509, Ranipet District, Tamil Nadu, India
Gowri Ayyadurai	Department of Biotechnology, Srimad Andavan Arts and Science College, Tiruchirappalli, Tamil Nadu, India
Hino Fernando E.	ICAR- Krishi Vigyan Kendra, TNJFU, Nagapattinam, Tamil Nadu, India
Kohila Subathra Christy	Department of Chemistry, A.P.C. Mahalaxmi College for Women (Affiliated to Manonmaniam Sundaranar University), Tuticorin 628002, Tamil Nadu, India
Kavitha Parangusadoss	Department of Microbiology, Srimad Andavan Arts and Science College (Autonomous), Trichy 620005, Tamil Nadu, India
Krishnaveni Jaisangar	Department of Microbiology, Srimad Andavan Arts and Science College (Autonomous), Trichy 620005, Tamil Nadu, India
Koperuncholan Marimuthu	Department of Botany, Srimad Andavan arts and Science College, Tiruchirappalli, Tamil Nadu, India
K Gobalan	Department of Biotechnology, Jamal Mohamed College, Tiruchirappalli, Tamil Nadu, India
Khaleel Basha Sabjan	Department of Chemistry & Biochemistry, C. Abdul Hakeem College (Autonomous) Melvisharam 632509, Ranipet District, Tamil Nadu, India
M. Rajeshwaran	ICAR-Central Institute of Fisheries Education, Mumbai, Maharashtra 400061, India
Mohammed Meharoof	ICAR-Central Institute of Fisheries Education, Mumbai, Maharashtra 400061, India
M. I. Delighta Mano Joyce	Department of Zoology, Sadakathullah Appa College (Autonomous), Tirunelveli, Tamil Nadu, India
M. Vijayalakshmi	Department of Zoology, A.P.C.Mahalaxmi College for Women, Thoothukudi, Tamil Nadu, India
M. Paripooranaselvi	Department of Zoology, St. Mary's College (Autonomous), Thoothukudi-628001, Tamil Nadu, India

P. J. Joslin	Department of Zoology, St. Mary's College (Autonomous), Thoothukudi, Tamil Nadu, India
Revathi Nandhakumar	Department of Biotechnology, Srimad Andavan Arts and Science College, Tiruchirappalli, Tamil Nadu, India
Reshma Devi Ramesh	Department of Biotechnology, Srimad Andavan Arts and Science College, Tiruchirappalli, Tamil Nadu, India
Sumithra Pasumalaiarasu	Department of Microbiology, Srimad Andavan Arts and Science College (Autonomous), Trichy 620005, Tamil Nadu, India
Sabareesvaran Kamaraj	Department of Microbiology, Srimad Andavan Arts and Science College (Autonomous), Trichy 620005, Tamil Nadu, India
Sudharsan Parthasarathy	Department of Biotechnology, Srimad Andavan Arts and Science College, Tiruchirappalli, Tamil Nadu, India
Siva Vijayakumar Tharmasivam	Department of Biotechnology, Srimad Andavan Arts and Science College, Tiruchirappalli, Tamil Nadu, India
Syed Muzammil Munawar	Department of Chemistry & Biochemistry, C. Abdul Hakeem College (Autonomous) Melvisharam 632509, Ranipet District, Tamil Nadu, India
Vajiha Banu Habeeb Mohamed	Department of Microbiology, Jamal Mohamed College, Trichy 620020, India

CHAPTER 1

Global Mariculture Production

C. Stella Packiam[1,*], **Dhivya Antony**[2], **Clara Jeya Geetha J.**[3] and **Kohila Subathra Christy**[1]

[1] *Department of Chemistry, A.P.C. Mahalaxmi College for Women, Thoothukudi, Tamil Nadu, India*

[2] *Department of Chemistry, Dhanalakshmi Srinivasan Arts and Science (Co-Ed) College (Affiliated to University of Madras), ECR, Mamallapuram, Chennai 603104, Tamil Nadu, India*

[3] *Department of Chemistry, Pope's College (Autonomous) (Affiliated to Manonmaniam Sundaranar University), Sawyerpuram 628251, Tamil Nadu, India*

Abstract: To date, the net contribution of global mariculture to the world's fish supplies, along with the changes in this contribution, has been briefly discussed in this chapter. Primarily, the industrial sub-sector emphasizes how mariculture is integrated into the global marine food system. However, the greatest contribution to global fish production and food security has come from inland aquaculture. Although the dependency on marine ingredients persists and the reliance on terrestrial ingredients has increased, significant improvements in aquaculture feed efficiency and fish nutrition have also been made. This has resulted in a decrease in the fish-in-fish-out ratio for all fed species. This ratio is expected to increase over the next few decades as global demand for seafood rises and fisheries catch remains stagnant. The marine and brackish subsector is of particular interest for examination, as the importance of mariculture increases due to its expanding impact on global mariculture growth, as well as its recognized detrimental effects on fish populations and maritime security. The ecological services provided by the cultivation of molluscs and seaweed are becoming more widely acknowledged. These services are being quantified, valued, and developed into a market to increase production.

Keywords: Fisheries, Global, Mariculture, Production.

INTRODUCTION

The output of aquatic flora and nutrients for fish around the world has increased dramatically over the last 50 years. Mariculture is a practice that produces food for human use. It is a process in which aquatic organisms, including both plants

* **Corresponding author Stella Packiam C.:** Department of Chemistry, A.P.C. Mahalaxmi College for Women, Thoothukudi, Tamil Nadu, India;
E-mail: stella.c6789@gmail.com

G. Chelladurai, M. Paripooranaselvi & B. Makesh Kumar (Eds.)

and animals, are cultivated in a controlled setting using an aquatic medium that may be entirely marine or in brackish water regions, or may be a mix of marine and freshwater. To encourage the standardization of mariculture data reports, this chapter is organized by grouping global mariculture production. Mariculture is generally considered to refer to the farming of aquatic animals and plants, such as fish, molluscs, crustaceans, and aquatic plants. To increase productivity, farming often involves some type of intervention, such as frequent stocking, feeding, and predator protection [1]. A record-breaking 214 million tonnes of fisheries and mariculture products were produced in 2020, including 178 million tonnes of aquatic animals and 36 million tonnes of algae, according to the 2022 edition of The State of World Fisheries and Mariculture (SOFIA). This growth was primarily attributed to Asia. Aquatic animal production was 30% greater in 2020 than it was on average during the 2000s and more than 60% higher than it was on average during the 1990s. These results were largely the result of the record mariculture output of 87.5 million tonnes of aquatic animals [2].

AQUATIC FOOD PRODUCTION AND CONSUMPTION

Food security and nutrition are being improved more than ever by aquatic foods. Aquatic food consumption has grown globally at an average annual rate of 3.0 percent since 1961, about twice the rate of annual population growth. Consumption has more than doubled since the 1960s, reaching 20.2 kg per person. Despite the COVID-19 pandemic's impact, over 157 million tonnes, or 89 percent of aquatic animal production, were used for direct human consumption in 2020. In 2019, aquatic foods accounted for roughly 17% of the animal proteins consumed, rising to 23% in lower-middle-income countries and exceeding 50% in certain regions of Asia and Africa. Fishing and mariculture support trading, employment, and economic growth. Aquatic animal production in 2020 was projected to generate a total first-sale value of $406 billion, with approximately $265 billion from mariculture [3]. The sector employed an estimated 58.5 million individuals, of whom about 21 percent were women, according to the most recent data. It is estimated that 600 million people rely on fisheries and mariculture in some capacity for their survival. To achieve equitable and sustainable development, it is essential to build resilience [4].

INTENTIONS BEHIND THE MARICULTURE

- To increase the quantity of marine fish produced.
- Utilize open-sea cage culture, shrimp hatcheries with greater diversity, and finfish seed production.
- Molluscan farming has broadened the field of marine agriculture.

- Popularising the idea of cage culture by creating demonstration models and units, and providing traditional fishermen with training.

ASSISTANCE-PROVIDING ELEMENTS

The following elements are supported by the National Fisheries Development Board (NFDB):

FINFISH SEED PRODUCTION IN SHRIMP HATCHERIES

For various finfish species, as well as significant crustacean species, breeding and seed production technologies have been developed in response to the increased emphasis on species diversity in recent years. To meet the demand of both farmers and the market, the research institutions of the Indian Council of Agricultural Research (ICAR) have developed breeding and seed production technologies for more than 60 species of freshwater, brackish water, and marine habitats over the past six decades. The species include highly prized ones with short production cycles, system- and region-specific demand, and many are rare or unheard of by the nation's mariculture industry, making them potential farmed fish of the future. To promote species diversification in fish farming and provide avenues for increasing the income of fish farmers in response to the Doubling Farmers' Income by 2022, these technologies are documented in the current book, “ICAR Technologies for Breeding and Seed Production of Finfishes and Shellfishes.” These technological briefings were compiled using data published in several ICAR Fisheries Research Institutes as research papers and other scientific reports. It is worth noting that the individuals listed as contributors for various species may not have developed the technologies, but rather participated in their demonstration [5].

CONSTRUCTION OF AN OPEN-WATER CAGE CULTURE

One of the four culture systems—ponds, raceways, recirculating systems, and cages—can be used to raise fish. A method that confines fish or shellfish in a mesh enclosure is called a cage or net pen. By definition, the structure of a cage and a net pen differs: a net pen has a rigid frame only at the top, whereas a cage has a rigid frame surrounding the entire structure. However, the terms “cage” and “net pen” are frequently used interchangeably [6]. Despite having rigid frames, marine cages are frequently referred to as net pens and vice versa. Differences in structure have minimal impact on how these systems produce things or how they affect the environment. Fish in cage culture are kept in some sort of mesh enclosure while still utilizing available water resources, such as ponds, rivers, estuaries, and the open ocean. Fish may be fed, observed, and harvested more easily since the mesh holds the fish in place [7]. Additionally, the mesh allows water to flow freely between the fish and the nearby water supply, thereby preserving good water quality and removing waste. Successful fish culture involves more than merely collecting a few fish at the end of the growing season, regardless of whether it is carried out in cages, open ponds, or in another way. The producer must be aware of the factors that affect success in order to have a successful and lucrative crop [8].

MODEL SEA CAGE CULTURE IMPLEMENTED FOR TRADITIONAL FISHERMEN

The depletion of fish stocks has led to a growing emphasis on mariculture within the fishing industry. The current trend shows that mariculture production has increased while the volume of wild harvest remains unchanged. In this case, cage farming plays a significant role in supplying the global market with fish products. It is one of the other sources for boosting mariculture output. The growth of this kind of fish production offers displaced and landless fishermen employment options as well as a long-term solution to the world's demand for fisheries goods. For underdeveloped coastal areas, caged fish farming is a lucrative industry [9]. Cage farming is a significant source of fish production and revenue for farmers, other industry stakeholders, and investors in several nations and regions.

Cage culture is now viewed as an alternative means of subsistence for those who have been relocated due to the building of reservoirs or the acquisition of land for other forms of development. Cage mariculture has emerged as a promising business in this context, enabling farmers to optimize the use of existing water resources, which are typically only partially utilized for other purposes. It was calculated that 2.2 million tonnes of mariculture were produced in Europe overall, utilising the cage culture technique. About 0.8 million tonnes of Atlantic salmon and about 80,000 tonnes of rainbow trout were produced in Northern Europe in 2004. The two fish species most frequently caught in the Mediterranean at present are European seabass and gilthead sea bream. Production has grown steadily over the past decade, with an average annual growth rate of 17%, increasing from 34,700 tonnes in 1995 to 137,000 tonnes in 2004. About 85% of the overall yield in 2004 came from these two species raised in cages. Over a couple lakh tonnes of salmonids were produced in cages from North and South America combined.

To minimize environmental impacts through increased dilution and potential visual pollution, cage culture systems must evolve. They can achieve this by moving farther offshore into deeper waters and more extreme operating conditions, or by integrating with lower-trophic-level species, such as seaweed, molluscs, and other benthic invertebrates. The idea behind co-cultivating lower trophic level species is that the waste outputs of some species groups, such as cage-reared finfish, can be utilized as inputs by other species groups, including seaweeds, filter-feeding molluscs, and/or benthic invertebrates like sea cucumbers, annelids, or echinoderms. Although considerable studies have been done utilising land-based systems, much more research is needed on open or offshore mariculture systems. Due to the chances for the integration of species and production methods near the coast as well as the potential for extension with cage placement farther from the coast, cage mariculture will be crucial to the process of producing enough (and palatable) fish for everybody. One of the main factors restricting the growth and expansion of mariculture is the ready availability of seed in commercial quantities. The only emerging industry to increase seafood output is marine cage farming, as the increasing exploitation strain on the wild stocks of many important marine fisheries has resulted in overexploitation and a decline in their catch [10].

MARICULTURE OF ORNAMENTAL FISH

The marine ornamental fish trade expanded from US$24 million to US$40 million annually in the 1980s to over US$300 million now. Worldwide, nearly 20 to 30 million marine reef fish are sold for consumption each year. Certain species are at risk as a result of the mounting pressure on natural stocks. The Sulawesi-native Banggai cardinalfish, *Pterapogon kauderni,* is a well-known example of the direct effects of fisheries on the aquarium trade. Banggai cardinalfish import costs decreased as supplies rose and the population fitness of the native population declined, resulting in an endangered condition on the IUCN Red List of Threatened Species (IUCN 2019).

The marine fish aquarium trade can be supported by captive breeding of marine ornamental fish species, which involves spawning, hatching, settlement, and growth to the juvenile or adult stage in an enclosed system [11]. To widely produce attractive marine fish, however, several crucial steps still need to be taken [12]. The rearing of larvae is one of the production bottlenecks for marine ornamental fish; according to our current understanding, many species produce larvae that are virtually impossible to maintain under suitable conditions, including adequate feeding. One option to overcome the current barriers to more captive-bred (CB) species being available is to enhance research efforts on marine ornamental fish mariculture. According to the CORAL list study, 338 species of marine ornamental fish have been successfully bred in captivity, with private businesses and knowledgeable hobbyists playing a significant role in this achievement. Although this figure has grown by an average of 8% since 2012, it still accounts for only 19% of the marine ornamental fish species traded for the aquarium hobby.

Furthermore, the small portion (7%) of these CB species is regularly sold on the market, including blennies, clownfish, and dottybacks. Other species, such as cardinalfish, goby, and some seahorses, are scarcely available on the market. In contrast, others, such as the CB surgeonfish, have only begun to be released into commerce in the past two years [13].

It is interesting to note that three of the top five market families, including Pomacentridae, Gobiidae, and Pomacanthidae, are among the most researched families with the greatest diversity of coral reef species. Eighty percent of the top ten best-selling species belong to the Pomacentridae family, which essentially dominates the aquarium business. In contrast to Pomacanthidae, which are among the most valuable species, Pomacentridae are typically low-value (per individual) species [14, 15]. It has been demonstrated that the worth of one species is highly correlated with its marketability. In light of this, we assume that the three distinct approaches listed below can adequately account for the similarities and contrasts between the patterns seen in academic research and the business sector:

- Private businesses would devote their R&D to highly valued species, such as Pomacanthidae, while concentrating on mass-producing tradeable and easily bred species, like Pomacentridae.
- Academics may work with a variety of species, depending on the situation, such as Chaetodontidae or Pomacanthidae species, which are challenging to breed but useful for studying specific research issues in repetitive studies.
- More experienced hobbyists would be less interested in rearing popular species in captivity and instead focus on novel species that have not yet been successfully raised in captivity on a large scale, without any commercial motivation.

The ability to supply the market with marine ornamental fish generated by closed-cycle mariculture is one of the objectives of ornamental fish mariculture. The risk in the market is that demand is not sustainably high enough in the long term to support the successful large-scale production of decorative marine CB fish. The good news for ornamental fish farming is that consumers are increasingly concerned about the sustainability of ornamental fish production, and price no longer seems to be a major deciding factor in their purchases. Mariculture is so beneficial that it may, in turn, encourage further study in this area [16]. It should also be noted that hobbyists' primary motivation for trading CB fish is, and will continue to be, the challenge of successfully reproducing species that are difficult to breed in captivity.

THE STATUS OF WORLD FISHERIES AND MARICULTURE 2022

Nearly 214 million tonnes of aquatic creatures are produced globally each year. The first-sale value of aquatic animal production was US$406 billion. Marine capture fisheries: 78.8 million tonnes, freshwater capture fisheries: 11.5 million tonnes. Animal mariculture production reached a new record of 87.5 million tonnes. Purchases and trade of marine organisms required for human consumption amount to 157 million tonnes. Moreover, a US$151 billion value of global trade in fisheries and mariculture products was achieved. The workforce and fleets of the 58.5 million people employed in the primary fisheries and mariculture sector. Asia has the most fish farmers and fishermen (84 percent). There are 4.1 million fishing vessels in the world. Asia has the largest fleet globally (2.68 million vessels, about two-thirds of the global fleet). Aquatic animal output was expected to have decreased slightly from the all-time high of 179 million tonnes in 2018 to 178 million tonnes globally in 2020. Ninety million tonnes (51%) of the contribution came from capture fisheries, while 88 million tonnes came from mariculture (49 percent). Seventy percent of the production (112 million tonnes) was gathered in coastal waters (from capture fisheries and mariculture, respectively), and thirty percent (66 million tonnes) was harvested in inland waterways (83 percent from mariculture and 17 percent from capture fisheries). The expected initial selling value of the entire global production, which includes USD 141 billion for capture fisheries and USD 265 billion for mariculture, was USD 406 billion [17].

THE NEED FOR FISHERIES AND MARICULTURE PRODUCTION

In its most recent "Food Outlook" report, the Food and Agriculture Organization of the United Nations (FAO) anticipates a 1.5% increase in worldwide fisheries and mariculture production, reaching 184.6 million tonnes in 2022. According to the FAO's predictions, the sector's growth increased by 92.2 million tonnes, or 2.9 percent, compared to the previous year. Along with pricing analyses and policy information, the report contains projections of the supply and demand for food commodities, as well as fish and fisheries products. The special section of this study places particular emphasis on current trends in the global food trade. Orientation to fostering a sustainable mariculture industry.

While sustainability issues persist in mariculture and fisheries, these industries remain crucial in ensuring food security and preserving livelihoods. The FAO is committed to continuing to support its Members in establishing a sustainable, inclusive, resilient, and empowering fisheries and mariculture sector. The use and processing of fisheries and mariculture products have undergone significant changes in recent years [18]. In contrast to the 1960s, when only 67 percent of global production was used for direct human consumption, 89 percent was used in 2020. The remainder was utilized for non-food purposes, with the majority going to fishmeal and fish oil, and the balance being allocated to pet food, ornamental fish, bait, pharmaceutical uses, livestock, fur production, and direct feeding in mariculture. The majority of aquatic food for direct human consumption was still in live, fresh, or chilled forms, with frozen, prepared, preserved, and cured following. The proportion of aquatic food output preserved by salting, smoking, fermenting, or drying is higher in Asia and Africa than it is globally.

MONITORING ORGANISATION IN MARICULTURE

The first gathering of the world's foremost platform for discussions and decisions on fisheries and mariculture trade since the start of the pandemic was attended by more than 300 delegates from 65 member countries and observers. At the end of 2019, the previous Sub-Committee meeting was held in Vigo, Spain. Consumer behavior has changed, and demand has adjusted accordingly. Fisheries and mariculture goods are produced and traded differently due to altered consumer behavior and demand changes resulting from supply problems for traditional fish species, increased transportation and fuel costs, and an overall rise in prices [19]. Still, it has the potential to improve social participation, reduce poverty, and enhance food security by opening up new markets that can pay small-scale producers, improve nutrition, and mitigate global inequalities. Moreover, women are quite well represented in this industry, particularly in processing jobs. Cohesive strategies in mariculture come to market, encompassing resource management, coastal management, mariculture, livelihoods, food security and nutrition, value chains, and aquatic food systems. The FAO works to implement integrated approaches. With a focus on operations that create capacity and establish standards, it collaborates with other international organisations to leverage their strengths.

BLUE TRANSFORMATION IN MARICULTURE

The initiative is firmly rooted in the "Blue Transformation" Program Priority Area, which aims to foster more effective, inclusive, resilient, and sustainable aquatic food systems, as described in the FAO Strategic Framework 2022-2031.

The goal of Blue Transformation is to secure and sustainably maximise the contribution of aquatic (both marine and inland) food systems to food security, nutrition, and accessible healthy diets for all. Blue Transformation is a targeted effort in which agencies, nations, and dependent communities use existing and emerging knowledge, tools, and practices [20].

KEY POINTS OF THIS DOCUMENT OFFER

- A vision for the FAO's work on aquatic food systems for the years 2022-2030 by outlining a pathway for the transformation of aquatic food systems called the "Blue Transformation."
- The 2021 Declaration for Sustainable Fisheries and Mariculture of the Committee on Fisheries of the Food and Agriculture Organization of the United Nations (FAO) and the FAO Strategic Framework 2022-2031 align with this Blue Transformation Roadmap.
- The optimized components of aquatic food systems' contribution to the Sustainable Development Goals (SDGs).
- Acknowledges the necessity of advancing to more effective, inclusive, resilient, and sustainable aquatic food systems in order to promote the 2030 Agenda and improve food production, nutrition, the environment, and quality of life for all, without leaving anyone behind.
- A broad framework to assist the operations of the FAO Fisheries and Mariculture Division (NFI), including planning, implementing, monitoring, and communicating.
- The significance of aquatic food systems as engines of employment, economic expansion, social progress, and environmental restoration all of which are foundational elements of the SDGs.

The FAO claims that more focused, transformative changes are required as the sector continues to grow, in order to produce a more sustainable, inclusive, and equitable fisheries and mariculture sector. The blue transformation drive outlined in the report seeks to alter the way mariculture is produced by advancing the development and adoption of sustainable mariculture practices.

- Incorporating mariculture into national, regional, and international development strategies, and changing the way that food is produced.
- Boosting inclusive livelihoods and increasing mariculture production to meet the rising demand for aquatic food.
- Increasing the capacity of the mariculture industry at all levels to create and use cutting-edge technology and management techniques.
- Fisheries and mariculture growth is crucial in our efforts to eliminate global hunger and malnutrition; however, further transformation is needed in the sector to address the challenges.
- To ensure that aquatic foods are obtained responsibly, livelihoods are protected, and aquatic habitats and biodiversity are preserved, we must restructure agrifood systems.

The Focus on Blue Revolution emphasizes the creation of an environment that supports the integrated and comprehensive management of fisheries, promoting the socioeconomic advancement of fishermen and fish farmers.

FISHERIES AND MARICULTURE INFRASTRUCTURE DEVELOPMENT FUND (FIDF)

The establishment of the Fisheries and Mariculture Infrastructure Development Fund (FIDF) in 2018-19 is part of the Indian government's efforts to improve the fishing industry, as outlined in the Union Budget for 2018. As a result, the Department of Fisheries, Ministry of Fisheries, Animal Husbandry & Dairying, Government of India, established the FIDF, a special fund with a total fund size of Rs 7522.48 crore, during the 2018-19 fiscal year. To achieve the goal of 15 million tonnes by 2020, established under the Blue Revolution, FIDF envisions building fisheries infrastructure facilities in both the marine and inland fisheries sectors, as well as increasing fish production. Additionally, the FIDF aims to achieve a sustainable growth rate of 8 to 9 percent, thereby increasing the nation's fish production to approximately 20 million tonnes by 2022-2023. Through Nodal Loaning Entities (NLEs), such as state governments, Union Territories, and State entities, FIDF offers concessional financing to Eligible Entities (EEs) for the development of specified fisheries infrastructure projects [21].

- All scheduled banks, National Cooperatives Development Corporation, and National Bank for Agriculture and Rural Development (NABARD).
- For providing concessional financing by the NLEs at an interest rate not lower than 5% per annum, the Department of Fisheries offers an interest subvention of up to 3% per annum under the FIDF. The FIDF loan term is five years, from 2018-19 to 2022-23, with a maximum repayment length of 12 years, including a 2-year principal repayment moratorium.

The nodal implementing agency for the FIDF Scheme is the National Fisheries Development Board (NFDB), located in Hyderabad. A FIDF Portal has been created online by NFDB. The portal provides information about the FIDF for all interested parties, an online application form for various entities, as well as online processing and approval of FIDF applications. For the sake of processing each application, each stakeholder has received a unique login. The portal also features a FIDF Dashboard to track the status of applications in real-time.

MANTRI PRADHAN MANTRI MATSYA SAMPADA YOJANA

The Government of India launched PMMSY as a component of the "Atma Nirbhar Bharat" package with an investment of Rs. 20,050 crores, the largest ever in the fishing industry. Fishermen are provided with access to financial aid, insurance coverage, and the Kisan Credit Card service to help them obtain the working capital they need. By 2024-2025, the initiative hopes to produce 22 million tonnes of fish. Approximately 55 lakh people are expected to benefit from the improved work prospects [22].

AIM & OBJECTIVES

By leveraging rural resources and rapidly strengthening the rural economy, PMMSY aims to achieve the goal of rural development. In the fisheries industry, PMMSY's primary tagline is "Reform, Perform and Transform."

The PMMSY scheme's improvements and initiatives have been ingrained in:

- Development of the core and trunk infrastructure.
- Modernization of Indian fisheries through initiatives like.
- Advocate for more fishing ports and landing areas.
- Traditional fishing boats, such as trawlers, are being modernized and mechanized.
- Reduce post-harvest loss by providing post-harvest facilities.
- Facilities for cold chains.
- Pristine and sanitary fish markets.
- Ice boxes on two wheels.

Achievements

From 2019-20 to 2021-2022, the fishing industry experienced an astounding 14.3% growth.

- A record-breaking 141.64 lakh tonnes of fish were produced in 2019-20, and that number increased to 161.87 lakh tonnes in 2021-22.
- Shrimp exports accounted for the majority of the 13.64 lakh tonnes worth of exports in the fishing industry, valued at Rs 57,587 crore.
- Exports are currently being made to 123 nations, including China, Thailand, Japan, Taiwan, Tunisia, the US, Hong Kong, Kuwait, *etc.*
- 31.47 lakh farmers have received support from PMMSY through insurance coverage from 22 states and 7 UTs.

Implementation

It is done through the use of an umbrella scheme with two distinct parts, namely:

Central Sector Scheme: The central government will cover the project's costs.

Centrally Sponsored Scheme: The States/UTs will undertake all supporting components and activities, and the Centre and the State will share the associated costs.

Prospective Strategy

The promotion of mariculture would be prioritized in Northern India's saline and alkaline regions.

A coordinated laboratory network will support the focus on aquatic health management, addressing issues related to disease, antibiotics, and residues.

OPTIONS TO ACCEPT

The foundation of PMMSY consists of fisheries and fish growers. By using technology, public stocking, and programmes to restore water bodies through river and sea ranching, we can harness the full potential of our reservoirs and natural resources.

For India to become the most productive nation in the world, scientific methods must be used in fish farming.

STRATEGIES IN MARICULTURE

Farming suggests that a person or organisation owns the stock being cultivated. For statistical purposes, mariculture refers to the harvest of aquatic organisms by an individual or corporate entity that has owned them throughout their entire rearing period. In contrast, fisheries refer to the harvest of aquatic organisms that are exploitable by the general public as a common property resource, with or without the necessary licenses. It is essential to note that the explanation provided refers to mariculture as a technique, rather than to what should be reported. Hatchery production, which is frequently quantified, relates specifically to seed production from indoor or outdoor hatchery/nursery facilities [1].

Mariculture output intended for final harvest for human consumption is referred to as mariculture production. The Yearbook reports mariculture production across three different culture environments: freshwater, brackish water, and marine water. Waters with consistently minimal salinity are classified as freshwater.

Waters that are salinity-appreciable but not consistently high are referred to as brackish water. It is typically distinguished by consistent daily and seasonal changes in salinity brought on by influxes of freshwater and fully potent marine water. The term "brackish" also refers to enclosed coastal and inland bodies of water where the salinity is higher than freshwater but lower than marine water. The term "marine" refers to coastal and offshore waters characterized by high salinity and minimal daily and seasonal fluctuations.

DATA ON FISHERIES PRODUCTION

The data on the volume and value of fisheries production for the year 2017 are presented in this Fisheries Situationer. It provides information on the state of the three major fisheries subsectors, which are commercial, municipal, and mariculture, by significant species. It is a product of the four fisheries surveys regularly conducted by the Philippine Statistics Authority (PSA): the Quarterly Commercial Fisheries Survey (QCFS), the Quarterly Municipal Fisheries Survey (QMFS), the Quarterly Inland Fisheries Survey (QIFS), and the Quarterly Aquaculture Survey (QAqS). The QCFS collects information on the volume of offloading at representative traditional subsector landing centres. The stratified simple random sampling approach was used to choose the sample landing centres. The data being collected includes the amount of cargo unloaded and the cost per kilogram of the top 31 species and additional species [19]. Additionally, information is gathered from the administrative files of non-traditional landing sites, including those operated by the Philippine Fisheries Development Authority (PFDA), private companies, and Local Government Units (LGUs).

OUTCOMING FORECASTS IN MARICULTURE

The analysis reveals the following key patterns during the years up to 2030: Production, consumption, and trade in global fisheries and mariculture are anticipated to rise, albeit at a slower rate over time. Due to higher output in regions where resources are adequately managed, the world's capture production is expected to increase significantly. Despite its slowing, global mariculture production is still expected to close the majority of the supply-demand gap. The aquatic food supply is expected to rise across the board. However, per capita consumption in Africa, particularly in sub-Saharan Africa, is predicted to decline slightly, raising concerns about the security of the continent's food supply.

The trade in aquatic products is expected to grow at a slower rate than in the previous decade. However, the proportion of production from fisheries and mariculture that is exported is anticipated to remain stable. Except for prices related to mariculture and trade, all prices are expected to rise in nominal terms but decline in real terms [18]. It is anticipated that the new fisheries and mariculture reforms and regulations that China will implement as part of its Thirteenth (2016-2020) and Fourteenth (2021-2025) Five-Year Plans will influence prices, output, and consumption globally. Furthermore, significant changes are projected to occur in the environment, resource availability, macroeconomic conditions, international trade regulations and tariffs, and market characteristics over the next ten years. These changes could have a medium-term impact on production, markets, and trade. Countries will become increasingly exposed to risks due to climate variability and change, including changes in the frequency and severity of extreme weather events [1]. These changes are also anticipated to have significant and geographically diverse effects on the availability, processing, and trade of aquatic products.

CONCLUSION

Mariculture is a well-established and rapidly expanding industry with the potential to provide millions more people with relatively low-impact, high-quality nutrition. However, it can also have long-lasting effects on marine and coastal habitats, as well as socio-ecological systems. Fine-scale, spatially explicit production data are essential for systematically and sustainably guiding mariculture production, minimizing competition with other human activities for land and water resources, and planning for the potential effects of climate change. While mariculturists successfully raise certain species, several desirable ones have not yet been economically produced. This lack of economic productivity is brought on by the complexity or difficulty of controlling their life cycles in culture.

Many people are concerned about the potential environmental effects of mariculture and oppose the presence of pens and cages in their bays. These and other challenges have been acknowledged, and scientists are working to address them. The objective is to produce high-quality seafood while remaining environmentally conscious. Our comprehensive simulation approach provides a previously unattained view of the global distribution of mariculture farms and could be further improved by incorporating more complex methods based on farm-level characteristics. To better inform policies and practices at all scales, coordinated efforts should focus on closing key data gaps and offering guidance for building a global, standardized mariculture location database.

REFERENCES

[1] Campbell B, Pauly D. Mariculture: A global analysis of production trends since 1950. Mar Policy 2013; 39: 94-100.
[http://dx.doi.org/10.1016/j.marpol.2012.10.009]

[2] Muhammed A. Oyinlola, Mariculture: perception and prospects under climate change, 22. Elsevier 2019; pp. 227-39.

[3] Food and Agriculture Organization of the United Nations (FAO). The State of World Fisheries and Aquaculture 2022. Rome: FAO; 2022.

[4] Issue Information. *Aquaculture, Fish and Fisheries*. 2022 Aug;2(4): 253-305.

[5] Sudhir Raizada, P. Pravin, J.K. Jena, ICAR Technologies: Breeding and seed production of finfishes and shellfishes, August 2019, Indian Council of Agricultural Research, New Delhi, India.

[6] SRAC Publication No. 163 - Cage Culture- Species Suitable for Cage Culture.

[7] SRAC Publication No. 164 - Cage Culture- Handling and Feeding Caged Fish.

[8] SRAC Publication No. 166 - Cage Culture- Harvesting and Economics.

[9] Shubhadeep Ghosh, Loveson Edward L, Uma Mahesh. V and Narasimhulu Sadhu, Regional Centre of ICAR-Central Marine Fisheries Research Institute, Visakhapatnam, Cage Culture of Marine Finfishes, Training Manual Series No. 10 / 2016.

[10] Cage Culture of Marine Finfishes. Training Manual Series No 10. Visakhapatnam, Andhra Pradesh, India: ICAR- Visakhapatnam Regional Centre of Central Marine Fisheries Research Institute 2016.

[11] Olivotto I, Chemello G, Vargas A, Randazzo B, Piccinetti CC, Carnevali O. Marine ornamental species culture: From the past to "Finding Dory". Gen Comp Endocrinol 2017; 245: 116-21. [PMID: 26969243]

[12] Rhyne AL, Tlusty MF, Szczebak JT. Early culture trials and an overview on U.S. marine ornamental species trade. In: Calado R, Olivotto I, Oliver PM, Holt J, Eds. Marine Ornamental Species Aquaculture. Hoboken, NJ: John Wiley & Sons Ltd 2017; pp. 51-70. a

[13] Rhyne AL, Tlusty MF, Szczebak JT, Holmberg RJ. Expanding our understanding of the trade in marine aquarium animals. PeerJ 2017; 5: e2949. b [PMID: 28149703]

[14] Wabnitz C, Taylor M, Green E, Razak T. From Ocean to Aquarium: The Global Trade in Marine Ornamental Species. Cambridge, UK: UNEP-WCMC 2003.

[15] Leal MC, Vaz MCM, Puga J, Rocha RJM, Brown C, Rosa R, *et al.* Marine ornamental fish imports in the European Union: an economic perspective. Fish Fish 2015; 17: 459-68.

[16] Moorhead JA, Zeng C. Development of captive breeding techniques for marine ornamental fish: a review. Rev Fish Sci 2010; 18: 315-43.

[17] Zaher SS, Ibrahim SM. Challenges and opportunities of bioplastics produced from algae. *Egyptian Journal of Aquatic Biology and Fisheries*. 2023 Sep 1;27(5):365-386.

[18] Costello C, Cao L, Gelcich S, Cisneros-Mata MÁ, Free CM, Froehlich HE, *et al.* The future of food from the sea. *Nature*. 2020 Dec 1;588(7836):95-100.

[19] PSA Websire: http://www.psa.gov.ph

[20] Available from: https://journalsofindia.com/blue-transformation-roadmap-2022-2030/

[21] Available from: www.fidf.in

[22] Available from: https://www.drishtiias.com/printpdf/pradhan-mantri-matsya-sampada-yojana-pmmsy

CHAPTER 2

Contemporary Status and Imminent Projections of Mariculture Advancements

Sumithra Pasumalaiarasu[1,*], Vajiha Banu Habeeb Mohamed[2], Kavitha Parangusadoss[1], Anathakumar Saminathan[1], Sabareesvaran Kamaraj[1] and **Krishnaveni Jaisangar[1]**

[1] *Department of Microbiology, Srimad Andavan Arts and Science College (Autonomous), Trichy 620005, Tamil Nadu, India*

[2] *Department of Microbiology, Jamal Mohamed College, Trichy 620020, Tamil Nadu, India*

Abstract: In the contemporary world, where the demand for proteins and minerals is rapidly increasing, the need for seafood is growing substantially. However, traditional fishing methods are insufficient to meet this demand, especially with a rising population. In this context, mariculture emerges as an innovative approach that can effectively address the limitations of wild fishing. Although highly efficient, mariculture also has certain challenges that need to be addressed. This chapter discusses the current status of mariculture, with a focus on India, and the technologies that have been implemented. Furthermore, it highlights the need for further development and governance to overcome these challenges and achieve efficient mariculture practices. Beyond serving as a source of human consumption, mariculture plays a vital role in the country's economy and employment.

Keywords: Nutrition, Population, Fishing, Technology, Economy, Employment.

INTRODUCTION

Mariculture is a branch of aquaculture that involves the cultivation of food products or other commercial products in the marine environment. It can be conducted under natural conditions or through human interventions, such as cages or sea enclosures. The main products obtained through mariculture include finfishes, crustaceans, molluscs, and other non-food products, such as jewelry, cosmetics, and pharmaceuticals. Furthermore, mariculture is considered an important contributor to India's economic growth and employment. Therefore, it requires active initiatives from both government and private sectors for its further

* **Corresponding author Sumithra Pasumalaiarasu:** Department of Microbiology, Srimad Andavan Arts and Science College (Autonomous), Trichy 620005, Tamil Nadu, India;
E-mail: sumithrachandramohanmb@gmail.com

G. Chelladurai, M. Paripooranaselvi & B. Makesh Kumar (Eds.)

development. The following chapter discusses in detail its significance, methods, current status, strategies, and advancements.

MARICULTURE, THE NEED OF THE HOUR

Global demand for nutrition is rising due to a growing population. Analyses have identified that approximately 2 billion people suffer from inadequate basic nutrition, and around 1 billion lack even the minimum calorie requirements [1]. In this context, protein is a fundamental nutritional requirement, which can be obtained from grains or edible meats [2]. Marine food products are rich in protein and essential micronutrients; however, they currently contribute only about 17% of global meat consumption. Projections estimate that seafood demand will increase by 36–74% by 2050 under current scenarios [3]. Therefore, the importance of marine food products is steadily increasing.

Apart from human consumption, marine products such as seaweeds, clams, oysters, and ornamental fishes play a vital role in the production of valuable products, including pharmaceuticals, bioactive compounds, fertilizers, drugs, cosmetics, biofuels, nutraceuticals, pearls, and ornamentals [4, 5]. Given these growing demands, improved production of marine products is essential, which cannot be met through natural resources alone. Therefore, innovative technologies must be introduced to address the insufficiency of marine products and their by-products effectively. This approach has led to the development of mariculture and the implementation of advanced technologies for its further advancement.

INDIA'S GLOBAL CONTRIBUTION TOWARDS MARICULTURE

Mariculture, the practice of culturing fishes, aquatic plants, and other aquatic animals, can be carried out either in natural marine environments or in inland waters such as rivers, estuaries, lagoons, canals, and reservoirs [6]. Asia tops the list in mariculture production, with around seven countries among the top ten leading producers [5]. In particular, India, the second-largest producer after China, is highly efficient in mariculture production due to its geographical location and abundant resources.

India, which ranks among the 12 megadiverse countries, is also well known for its richness in flora and fauna [7]. Specifically, regarding marine life, it ranks at the top, comprising 3,231 species of fishes from both marine and freshwater environments, representing 9.70% of the 33,059 fish species identified worldwide [8]. Of these, 2,443 species, nearly 75.6% of the total, are classified under 230 families [9].

Approximately 10% of the global fish diversity is found in India, due to its extensive water resources from Himalayan lakes, rivers, the deep sea [10], and the Western Ghats. More significantly, the central part of the Indian Ocean is advantageous as it connects three distinct marine regions: the Indian Ocean, the Arabian Sea, and the Bay of Bengal [7]. Furthermore, India's coastline, Exclusive Economic Zone (EEZ), and continental shelf, which cover approximately 8,129 km, 2.02 million km^2, and 0.5 million km^2 respectively, contribute to an impressive fisheries yield of 3.93 million tonnes [11]. According to Lakra and Gopalakrishnan (2021), this production has further increased globally to 13.4 million tonnes in 2018–2019, including 3.7 million tonnes from marine sources and 9.7 million tonnes from inland waters. At present, the most commonly cultured species in mariculture include fishes, crustaceans, molluscs, seaweeds, and some other less common species [5]. To produce these valuable products more efficiently, various initiatives have been implemented, both technologically and through governance, schemes, and policies. However, it is widely acknowledged that all fields of innovation and implementation have their own advantages and limitations, which must be addressed to further enhance the status of mariculture.

RECENT STATUS OF INDIA IN MARICULTURE

Recently, CMFRI (2020) reported an overview of the marine fisheries status for 2020, in comparison to 2019, which is further discussed in detail.

Fish Landings of Marine Fisheries

Fish landings reflect the growth and development status of fisheries and mariculture. "Marine fish landings" refers to the quantity of fish captured, post-harvest, and brought to either international or domestic ports. In 2020, Tamil Nadu led marine fish landings, contributing 20.51% (5.59 lakh tonnes) of the total. Gujarat and Karnataka ranked second and third, with 19.51% (5.32 lakh tonnes) and 13.75% (3.75 lakh tonnes), respectively. A decline of 23.45% was observed in 2020 (2.73 million tonnes) compared to 2019. The COVID-19 pandemic was a major factor in this reduction, as marketing, sales, and distribution were severely disrupted, and fresh fish were not easily available. Marine fish landings by coastal states and the number of species are summarized in Table **1**, comparing recent years.

Table 1. Status of marine fish landings, a state wise information.

S. No	Coastal States	State landings (Lakh tonnes)		Landings in percentage		Number of landings species	
		2019	2020	2019	2020	2019	2020
1.	Gujarat	7.49	5.32	21.04	19.51	241	199
2.	Daman and Diu	1.12	0.39	3.14	1.42	122	114
3.	Maharashtra	2.01	1.40	5.64	5.13	268	202
4.	Goa	0.33	0.59	0.92	2.10	168	131
5.	Karnataka	5.01	3.75	14.08	13.75	221	198
6.	Kerala	5.44	3.61	15.27	13.24	487	449
7.	West Bengal	2.49	2.60	6.99	9.54	276	229
8.	Odisha	1.02	1.75	2.87	6.43	294	269
9.	Andhra Pradesh	2.59	1.95	7.27	7.13	252	233
10.	Puducherry	0.37	0.34	1.04	1.24	262	232
11.	Tamil Nadu	7.75	5.59	21.75	20.51	612	594

Source [4]:

Dominating Species

In terms of species diversity, Tamil Nadu also ranks first, followed by Kerala and Odisha, for both 2019 and 2020. The dominant groups in India's marine fisheries include lesser sardines, Indian mackerel, penaeid shrimps, ribbonfishes, and threadfin breams. Notably, oil sardine, which was the top species in previous years, has declined to the 10th position.

Prevalent Resources

On the whole, the dominant resources include Pelagic, Demersal, Crustaceans, Molluscan, and Miscellaneous, which contribute 37%, 27%, 19%, 11%, and 6%, respectively. Based on geographical location, the quantity and variety of species change, which is much more important in resource management. The contributions of each state, based on its groups, are represented in Fig. (**1**).

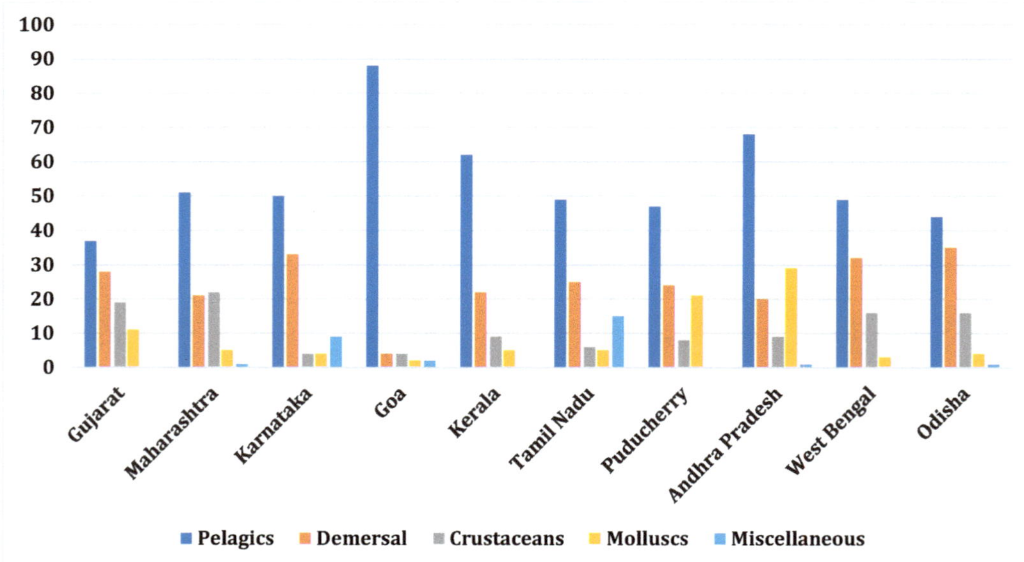

Fig. (1). State wise representation of dominating marine groups.

Similarly, some fish species are found to be abundant in certain states, as shown in Table **2**.

Table 2. Predominant fish varieties in varying states.

S. No	States	Pelagics	Demersal	Crustaceans	Molluscs
1.	**Gujarat**	• Ribbon fish • Bombay duck • Clupeids • Carangids • Tunas • Seer fishes • Indian mackerel	• Croakers • Catfish • Threadfin breams • Rock cod	• Non-penaeid shrimp • Penaeid shrimp • Crabs • Stomatopods	• Cuttlefish • Squids • Octopus
2.	**Maharashtra**	• Indian mackerel • Bombay duck • Lesser sardines • Horse mackerel	• Croakers • Pomfrets • Catfish	• Non-penaeid shrimp • Penaeid shrimp	• Squids • Cuttlefish

(Table 2) cont.....

S. No	States	Pelagics	Demersal	Crustaceans	Molluscs
3.	**Karnataka**	• Indian mackerel • Lesser sardines • Ribbon fish • Carangids • Tunas • Seer fishes • Half and full beaks	• Threadfin breams • Lizard fishes • Bullseyes • Rock cod • Croakers • Pomfrets • Silver bellies • Flat fishes • Elasmobranchs • White fish	• Penaeid shrimp • Crabs • Stomatopods	• Squids • Cuttlefish • Octopus
4.	**Goa**	• Indian mackerel • Carangids • Lesser sardines • Tunas • Horse mackerel • Seer fishes • Ribbon fish	• Flat fish • Lizard fish	• Penaeid shrimp • Crabs	• Squids
5.	**Kerala**	• Indian mackerel	• Threadfin breams • Lizard fish • Soles • Perches • Black pomfret • Rock cod	• Penaeid shrimp • Crabs	• Squids • Cuttlefish
6.	**Tamil Nadu**	• Lesser sardines • Carangids • Oil sardine • Mackerels • Tunas • Barracudas	• Silver bellies • Lesser perches • Pig face bream • Threadfin breams • Goat fishes • Croakers • Lizard fish	• Shrimps • Crabs • Lobsters	• Indian Ocean squid • Pharaoh cuttlefish • Cuttlefish • Octopus • Corals
7.	**Andhra Pradesh**	• Oil sardine • Indian mackerel • Lesser sardines • Ribbon fish	• Croakers • Silver bellies • Catfish	• Penaeid shrimp • Crabs • Non-penaeid shrimp	• Cuttlefish • Squids • Gastropods
8.	**West Bengal**	• Tuna • Half and full beaks	• Sciaenid • Pomfrets • Catfish • Soles	• Penaeid shrimp • Non-penaeid shrimp • Lobsters	• Cuttlefish

(Table 2) cont.....

S. No	States	Pelagics	Demersal	Crustaceans	Molluscs
9.	**Odisha**	• Carangids • Ribbon fish • Mackerels • Bombay duck • Seer fish • Tuna • Mullets • Half and full beaks • Barracudas	• Croakers • Pomfrets • Perches • Catfish • Silver bellies • Elasmobranch • Threadfin breams • Eels • Goat fish	• Non-penaeid shrimp • Penaeid shrimp • Crabs • Lobsters	• Squids • Octopus • Cuttlefish

Source [4]:

MAJOR TYPES OF FARMING

Mariculture involves varying amounts of cultivation, but some groups are more predominant and provide a higher yield.

Shrimp Farming

During the early 1970s, the importance and value of shrimp were recognized, which led to the development of breeding methods and rearing practices. Outstanding species, such as *P. indicus*, *P. monodon*, *M. monoceros, M. affinis*, M. dobsoni, and P. stylifera, were cultivated to a greater extent, with P. monodon showing an unrecognizable yield. A major threat emerged in 1994 with the outbreak of White Spot Syndrome Virus (WSSV), causing significant losses. As an alternative, *Litopenaeus vannamei*, a resistant and closely monitored species, was introduced. In response to these challenges, India adopted open-sea cage farming methods, while also recognizing the limitations of monoculture [12].

Seaweed Farming

Seaweeds are not widely consumed in India, yet they possess valuable phycocolloid-producing properties. They are commonly farmed using bamboo rafts, which yields good results [13]. Additionally, net tube cultivation of seaweeds, combined with lobster and finfish farming, has proven to be effective, cost-efficient, and superior in terms of quality. *Kappaphycus alvarezii* is well-established as a cultivable species, while other species are still under trial-and-error cultivation [12].

Lobster Farming

It has increased demand according to international standards, but due to a lack of hatching facilities, it has not gained significant importance. *Panulirus polyphagus*, the predominant lobster, is cultivated using pits, which are also highly valuable [11].

Crab Farming

It has emerged as an alternative to shrimp farming due to its resilience against viral disease-related mortality. *S. tranquebarica* is an economically viable species, cultivated either through monoculture or polyculture methods alongside milkfish and mullets on a small scale [11].

Oyster Farming

Pinctada fucata significantly contributes to pearl production and is cultured using raft and rack methods. Shell beads are implanted into the oyster shell, which subsequently develops into a pearl. *In vitro* pearl production from mantle tissue culture is currently under research [11]. Regarding pearl farming training, entrepreneurs signed a Memorandum of Understanding in 1996.

Ornamental farming

The Indian Ocean is a diverse land consisting of coral reefs and ornamental fish, including clownfish, damselfish, and seahorses. These ornamental fish are produced using hatchery technologies that increase the country's foreign exchange [11].

TECHNIQUES INVOLVED

Many varieties of techniques and instruments are utilised in mariculture, based on the purpose. Certain contemporary techniques that work well as of today include:

Fishing Fleet

Fishing fleets refer to large vessels or boats used for fishing. It includes crafts (9%), motorized crafts (26%), and mechanized boats (65%) in the yield of marine fisheries [14]. According to CMFRI, mechanized fishing vessels are more efficient, yielding 83% of the total, followed by motorized fishing crafts (16.1%) and non-mechanized crafts (1%). This efficiency, of course, varies based on the species obtained and the location of capture. Although mechanized fishing is much more efficient, it is not very prevalent due to a lack of awareness among farmers.

Gears

Nets and trawlers are classified as fishing gears, representing a crucial component of mariculture practices. Bag nets, gill nets, and drift nets are commonly used across most coastal regions, while purse seines, seines, and mechanized gill nets are predominant in the southern areas. Multi-day trawlers have been found to be more efficient than other gill nets and seines [4].

Cage farming

Cage culturing of fisheries involves raising aquatic organisms in cages placed in the marine environment, providing a semi-natural habitat. It is one of the most common and widely practiced methods of mariculture. Oyster and crab farming are among the primary applications of cage culture. To address issues such as rusting of galvanized sea cages, fiberglass-reinforced plastic coatings were introduced, which also facilitate small-scale farming [12].

EXCEEDING THE SIGNIFICANCE OF MARICULTURE AND ITS CURRENT POSITION

Mariculture was introduced in India during the 1960s by the renowned ICAR-Central Marine Fisheries Research Institute (CMFRI), which implemented scientific tools and technologies, creating a breakthrough in the field of fisheries [12]. This led to several positive outcomes, such as:

Nutritional Resource

An analysis has identified that if fish production does not increase to meet demand, 10% of the global population will suffer from a lack of iron, zinc, and Vitamin A, while 20% will be deficient in most micronutrients [15]. In this context, mariculture provides a valuable solution, as marine edible products are highly nutritious, containing abundant proteins and micronutrients [1]. More importantly, they supply essential vitamins and minerals, including zinc, iron, fatty acids, Vitamin A, and Vitamin D [15]. Larsen *et al.* (2011) stated that the fatty acids present in seafood play a crucial role in brain development, prevent cardiovascular diseases, and support metabolism [16]. In addition, children and pregnant women particularly benefit from the nutrition provided by seafood, as it promotes better growth and development [15]. This nutritional necessity drives the development of mariculture.

Economically Valuable

Fisheries play a major role in trade for many developing coastal nations as well as developed countries, as they rely on foreign exchange for income [2]. India, in particular, possesses remarkable species richness, with 89 families including 548 species considered highly valuable commercial products [17]. In the year 2000–2001 alone, marine food exports generated an income of Rs. 63 thousand million [11]. This biodiversity and its commercial value significantly boost the country's economy. Mariculture contributes substantially to the total Gross Domestic Product (GDP) of India, accounting for 1.03% overall and 6.58% within the agricultural sector, generating around USD 6.73 billion in 2018–2019 [6]. Based on this revenue and profit, mariculture is considered a key component of the Indian economy.

Employment

Mariculture serves as an important source of employment in rural coastal areas, especially in developing countries like India, where more than 2 million people work in fish capture and related activities [6]. Additionally, due to the high demand for exports, post-harvest fisheries activities play a vital role in the labor force, further increasing employment opportunities [11]. Shrimp farming, in particular, has effectively increased employment rates by 2–15%, which in turn improved laborers' incomes by 6–22%. Similarly, seaweed farming [12], pearl farming [11], and ornamental fish culture are sectors where women are actively involved [4]. Another area with significant female participation is the selling of dry fish, which has received less attention [4]. This involvement of women not only provides them with income and employment but also empowers them through greater economic and social participation.

HINDRANCES FACED IN MARICULTURE

As of today, more than 100 countries depend on Indian seafood products [12], with over 80 million metric tonnes of fisheries and other marine products obtained from the oceanic environment [18]. Although mariculture offers significant advantages, certain constraints remain, which must be addressed to achieve effective and sustainable results.

Exploitation

According to India's fishing strategies, coastal areas up to a depth of 100 metres are heavily harvested, with more than 90% of the catch obtained within 50 metres, resulting in overfishing that exceeds the Maximum Sustainable Yield (MSY) limits [6]. Serajuddin *et al.* (2018) report that 31% of the overall fish stock is categorized as overfished, while 58% is fully exploited and cannot be replenished. Irresponsible fishing practices, such as dynamiting and poisoning, lead to mass fish mortality [7]. Therefore, effective conservation strategies need to be implemented to address these issues.

Environmental Factors

Rising sea surface temperatures have a significant impact on fisheries development, and are expected to increase by 2.0 to 3.5°C by the end of this century in the Indian Ocean. Ocean warming leads to decreased dissolved oxygen and disrupts nutrient and carbon cycling [6]. In addition, other concerns such as cyclones, ocean acidification, greenhouse gas effects, and coral bleaching contribute to fish depletion [7].

Anthropogenic Activities

Pollution is a major factor in the depletion of marine life, including plastic waste, oil spills, chemicals, and industrial sewage. These pollutants adversely affect marine organisms, sometimes causing genetic-level alterations [19]. Oil spills can increase mortality rates or disrupt spawning patterns, thereby reducing the stock of marine resources [7].

Due to these concerns, marine fisheries are operating at suboptimal yield levels, emphasizing the need for conservation technologies to support sustainable mariculture development.

CONCERNS OF THE DECLINE AND CONSERVATION STRATEGIES

Depending on the decline noted in various species, certain management steps need to be undertaken and anticipated to avoid such issues in the near future.

Pelagic Fishes

Pelagic fishes refer to the fishes that inhabit the pelagic zone, which is neither closer to the shore nor to the sea bottom. They are classified into coastal pelagic and oceanic pelagic based on the depth at which they live. Coastal pelagic species are those that are present above the continental shelf, whereas oceanic pelagic species live under the continental shelf.

According to CMFRI, pelagic fishes account for the majority of India's fish landings. Although large pelagics are caught using a combination of crafts and gears, the most common methods—contributing over 83%—include trawls, gillnets, and seines. The dominant species of large pelagics, based on harvest ratios, include barracudas, billfishes, cobia, dolphinfish, needlefish, queenfish, rainbow runners, seerfish, and tunas. Among these, tuna, seerfish, barracuda, and queenfish are identified as being below their exploitation potential on average annually. This has implications for fisheries landings, as most exports depend primarily on tunas and seerfish. Therefore, certain technical measures are required, such as deploying vessels in the open seas, although financial constraints remain a hindrance.

Bivalves

Bivalves are molluscs that occur in both marine and freshwater environments, consisting of two shells enclosing a soft body. In India, oysters, clams, and mussels are the leading bivalves produced, prevalent in Kerala, Odisha, Karnataka, Goa, Maharashtra, Tamil Nadu, and Andhra Pradesh. Clams represent the major contribution, followed by oysters and mussels. In 2020, Tamil Nadu's contribution decreased by 20%, which is a serious concern. This reduction has led to increased demand, resulting in the overexploitation of certain species. Therefore, adhering to proper spawning seasons is essential to mitigate such issues.

Gastropods

Univalves are a group of organisms that have a single shell enclosing a soft body. They can be found in saline water, freshwater, or on land. They are classified as invertebrates under the phylum Mollusca. Tamil Nadu contributed nearly 95% of the gastropod harvest. Capture methods include skin diving, gill nets, and trawl nets.

Gastropod catch by skin diving in 2020 resulted in 872 tonnes, with the dominant species being *C. ramosus* (67%), *T. pyrum* (29%), and *L. lambis* (5%). In gill net capture, the nets are set at the bottom to yield the best catch, producing about 225 tonnes of gastropods, including *T. pyrum, C. ramosus, L. lambis*, and *Babylonia spp.*, in that order of yield. Mechanized trawlers yielded up to 640 tonnes, comprising 22% strombids, 10% muricids, and 10% naticids.

ELASMOBRANCHS

Elasmobranchs are cartilaginous fishes, including sharks, rays, and related species. The pandemic significantly impacted their landings, causing a 40% decrease. Fishing is typically carried out using mechanized, motorized, and non-motorized sectors, with trawl nets in the mechanized sector and gill nets proving to be the most effective. A major concern within this group is the increased occurrence of deformities, which warrants further research. In a collaborative project on sharks and rays, a survey based on questionnaires was conducted among farmers and along the consumer chain. Improvements are also needed in export practices, trade statistics, and stakeholder awareness.

RETRIEVAL OF THE DEPLETED STOCK

According to the International Union for the Conservation of Nature and Natural Resources (IUCN), many species in the Indian coastal regions are on the verge of extinction or are endangered. This scenario needs to be reverted to increase mariculture production. It includes two major methods: *in situ* or *ex situ.*

i. *In situ* refers to the conservation of marine species within a protected zone, such as Marine Protected Areas (MPAs). It may be either wildlife sanctuaries, national parks, or biospheres.
ii. The *ex situ* method involves conservation in either laboratories or an artificial environment. It may include methods such as pond culturing, tank farming, or other techniques, such as gene-level modifications [7].

Restocking

It is possible only in an environment where other species do not dominate the targeted one, and moreover, their negative effects are negligible. This method is implemented to recover lost species on a large scale, creating a greater impact at the gene level [7]. Based on this, Ryman *et al.* (1995) suggest two concerns that;

i. Interbreeding results in the loss of native genetic specification upon the adaptation of novel genes.
ii. Diseases could destroy the total population [20].

Enhancement of Stock

Stock enhancement is primarily practiced to increase the population of a species and boost production rates. The tendency of some species to inhabit shallow areas and exhibit sedentary behavior can lead to overpopulation in confined spaces, increasing mortality rates [21]. In recent years, marine stocking has been recognized as a scientific practice [22] and is now considered an integral component of fishery management [23].

Sea Ranching

In this process, the cultured organisms are introduced into maintained environments, which include the culturing sites. During the post-incubation period, the grown-up organisms are harvested. Many countries, including the USA, Norway, and Japan, have used this method to improve stock [7].

CURRENTLY AVAILABLE TOOLS

i. Coastal and Marine Spatial Planning (MSP) – a scientific tool that addresses the problems of oceanic management and aims at conservation of marine culture and economic development [12].
ii. Geographical Information System – that identifies the marine sites, potential for farming, and provides decision support [24].
iii. National Marine Living Resources Data Centre (NMLRDC) – Database that provides information regarding the landings of different marine species.
iv. Satellite Remote Sensing (SRS) – used in establishing infrastructure inside mariculture sectors.
v. Geographic Information System (GIS) – indicates the locality of various purposes, such as reserved areas and rates of fish mortality [7].
vi. Environmental *impact assessment* (EIA) – a tool that helps identify environmental issues [5].
vii. Global Fishing Watch – collects information regarding the global footprint [25], financial aspects of fishing [26], and acts of illegal fishing [27].

GOVERNANCE AND ITS STATUS

Governments' support and ideas are important for the development of any technological and scientific innovations. Similarly, in mariculture, various institutions and organizations are involved that provide support in either or both ways.

Organisations and Institutions

Such renowned institutions include:

i. ICAR-Central Marine Fisheries Research Institute (CMFRI)
ii. ICAR-Central Institute of Brackishwater Aquaculture (CIBA)
iii. Rajiv Gandhi Centre for Aquaculture (RGCA)
iv. CSIR-Central Salt and Marine Chemicals Research Institute (CSMCRI) [12]
v. Marine Products Export Development Authority (MPEDA)
vi. Ecosystem Approach to Fisheries Management (EAFM)

Schemes and Policies

Similarly, the schemes that are implemented by the Indian government include:

i. "Pradhan Mantri Matsya Sampada Yojana (PMMSY)" – An umbrella scheme exhibited by "Fisheries management framework and fishers' welfare",

Government of India, that bridges the gap between the production sector and the marketing sector, connecting the consumers.

ii. The Centrally Sponsored Scheme (CSS) on Blue Revolution provides financial assistance for the development of infrastructure, construction of technical sites, post-harvest facilities, and transportation facilities, among others.
iii. The National Mariculture Policy-2019 (NMP) aims to promote the sustainable utilization of available resources, thereby increasing the country's nutritional security [12].
iv. Marine Fisheries Regulation Acts (MFRA) – that controls the gears and their utilization in the specified areas, along with norms regarding the instrumentation.
v. Deep Sea Fishing Policy (DSFP) – which provided accessibility of the Exclusive Economic Zone of India.
vi. The National Inland Fisheries and Aquaculture Policy (NIFAP) 2019 aims to develop the sustainable development of aquaculture and inland fisheries, which also includes brackish water and saline water [6].

CURRENT TECHNOLOGIES

Though new methods are implemented in marine fisheries, there are still some developments needed in the existing technologies:

i. High-density polyethylene (HDPE) rafts are extensively used in seaweed cultivation as they are allowed to float on the surface in various shapes.
ii. Sea cage farming gains more importance, especially in the field of capture-based farming, which includes lobsters.
iii. Monoculture of orange-spotted grouper produced increased yields when pellet feed was fed and low fish densities were provided.
iv. Co-culturing orange-spotted grouper in cages with Indian pompano resulted in a good production ratio.
v. Integrated multitrophic aquaculture (IMTA) that involves the integration of two species (*i.e.*, seabass and pearl spot in a cage that includes seaweeds and mussels. Mussel growth was extremely outstanding [4].

FUTURE RECOMMENDATIONS

Measures have been taken to develop mariculture and intensify its status both economically and environmentally. However, despite certain constraints, there is still a need for further development. A few of the recommendations are listed:

i. Recirculating Aquaculture Systems (RAS) should be given importance to expand horizontal and vertical intensive farming.

ii. Integrated Multi-Trophic Aquaculture (IMTA) produces a greater yield, but it is still in its infancy and needs to be developed.
iii. Awareness should be provided to the traditional fishers so that resource conservation can be increased.
iv. Breeding and mass seed production strategies should be enhanced to maximize the potential yield of fisheries.
v. Experts should actively be involved in the farm monitoring of fish.
vi. High-end research technologies should be utilized to develop an environmentally safe mariculture method.
vii. Collaborations should be initiated from both the National and International sectors.
viii. Academic institutions should provide students with the required knowledge about fisheries and its importance to take it to the next level [6].
ix. Entrepreneurs from both small-scale and large-scale enterprises should be encouraged and supported, especially through funding.

CONCLUSION

Mariculture is likely to develop significantly due to its abundant resources, biodiversity, nutritional value, and economic growth, among other factors, which also require proper management and conservation strategies. Currently, India is well-positioned in terms of mariculture due to its abundant resources. Its further development and future depend on recent technical, technological, scientific, and skilled workers. Implementation of proper regulations and policies, followed by all, will definitely create an outstanding global status for India.

REFERENCES

[1] Liu OR, Molina R, Wilson M, Halpern BS. Global opportunities for mariculture development to promote human nutrition. PeerJ 2018; 6e4733 [http://dx.doi.org/10.7717/peerj.4733] [PMID: 29761051]

[2] Sanchirico JN, Wilen JE. Global marine fisheries resources: status and prospects. Int J Glob Environ Issues 2007; 7(2/3): 106-18. [http://dx.doi.org/10.1504/IJGENVI.2007.013569]

[3] Costello C, Cao L, Gelcich S, *et al.* The future of food from the sea. Nature 2020; 588(7836): 95-100. [http://dx.doi.org/10.1038/s41586-020-2616-y] [PMID: 32814903]

[4] Cabral RB, Mayorga J, Clemence M, *et al.* Rapid and lasting gains from solving illegal fishing. Nat. Ecol. Evol. 2018; 2:650.

[5] Phillips M. Mariculture Overview. Encyclopedia of Ocean Sciences (Second Edition) 2009; 537-544 [http://dx.doi.org/10.1016/B978-012374473-9.00752-9]

[6] Lakra WS, Gopalakrishnan A. Blue revolution in India: Status and future perspectives. Indian J Fish 2021; 68(1): 137-50. [http://dx.doi.org/10.21077/ijf.2021.68.1.109283-19]

[7] Serajuddin M, Bano F, Awasthi M, Gupta P, Kumar G. Serajuddin M, Bano F, Awasthi M, Gupta P, Kumar G. Marine stock enhancement in India: Current status and future prospects. IntechOpen; 2018

Aug 1

[8] Eschmeyer WN, Fong JD. 2014.Catalog of fishes http://research. calacademy. org/research/ichthyology/catalog

[9] Gopi KC, Mishra SS. Diversity of marine fish of India InMarine faunal diversity in India 2015. [http://dx.doi.org/10.1016/B978-0-12-801948-1.00012-4]

[10] Lal KK, Jena JK. Fish genetic resources - India. In: Tyagi RK, Munasinghe DHN, Deepananda A, Frank Niranjan KHM and Khetarpal RK (Eds.), Regional Workshop on Underutilized fish and marine genetic resources and their amelioration - Proceedings and recommendations, Asia-Pacific Association of Agricultural Research Institutions (APAARI), Bangkok, Thailand 2021; 10-12: 55.

[11] Pillai NG, Mohan Joseph M, Ganga U. Marine fishing practices and coastal aquaculture technologies in India. A profile of people, technologies and policies in Fisheries sector in India. 2003:83-122.

[12] Mojjada SK, Divu D, Johnson B, Imelda J, Gopalakrishnan A. Mariculture advancemnts in India: Towards a new epoch. Infofish Int 2021; 55-8.

[13] Johnson J, Ignatius B. Seaweed farming in India progress and prospects. Indian Farming 2020; 70(11)

[14] Shinoj P, Gopalakrishnan A, Jena JK. Demographic change in marine fishing communities in India

[15] Golden CD, Allison EH, Cheung WWL, *et al.* Nutrition: Fall in fish catch threatens human health. Nature 2016; 534(7607): 317-20. [http://dx.doi.org/10.1038/534317a] [PMID: 27306172]

[16] Larsen R, Eilertsen KE, Elvevoll EO. Health benefits of marine foods and ingredients. Biotechnol Adv 2011; 29(5): 508-18. [http://dx.doi.org/10.1016/j.biotechadv.2011.05.017] [PMID: 21683131]

[17] Talwar PK, Kacker RK. Commercial sea fishes of India. Zoological Survey of India 1984.

[18] Costello C, Ovando D. Status, Institutions, and Prospects for Global Capture Fisheries. Annu Rev Environ Resour 2019; 44(1): 177-200. [http://dx.doi.org/10.1146/annurev-environ-101718-033310]

[19] Mustafa S, Ed. Genetics in sustainable fisheries management. 1999.

[20] Ryman N, Utter F, Laikre L. Protection of intraspecific biodiversity of exploited fishes. Rev Fish Biol Fish 1995; 5(4): 417-46. [http://dx.doi.org/10.1007/BF01103814]

[21] Bell JD, Munro JL, Nash WJ, *et al.* Advances in Marine Biology V49: Restocking and stock enhancement of marine invertebrate fisheries. Academic Press 2005.

[22] Bell JD, Leber KM, Blankenship HL, Loneragan NR, Masuda R. A new era for restocking, stock enhancement and sea ranching of coastal fisheries resources. Rev Fish Sci 2008; 16(1-3): 1-9. [http://dx.doi.org/10.1080/10641260701776951]

[23] Grati F, Scarcella G, Bolognini L, Fabi G. Releasing of the European sea bass Dicentrarchus labrax (Linnaeus) in the Adriatic Sea: Large-volume versus intensively cultured juveniles. J Exp Mar Biol Ecol 2011; 397(2): 144-52. [http://dx.doi.org/10.1016/j.jembe.2010.11.017]

[24] Divu DN, Mojjada SK, Pokkathappada AA, *et al.* Decision-making framework for identifying best suitable mariculture sites along north east coast of Arabian Sea, India: A preliminary GIS-MCE based modelling approach. J Clean Prod 2021; 284124760 [http://dx.doi.org/10.1016/j.jclepro.2020.124760]

[25] Kroodsma DA, Mayorga J, Hochberg T, *et al.* Tracking the global footprint of fisheries. Science 2018; 359(6378): 904-8. [http://dx.doi.org/10.1126/science.aao5646] [PMID: 29472481]

[26] Sala E, Mayorga J, Costello C, *et al.* The economics of fishing the high seas. Sci Adv 2018;

4(6)eaat2504
[http://dx.doi.org/10.1126/sciadv.aat2504] [PMID: 29881780]

[27] Cabral RB, Mayorga J, Clemence M, *et al.* Rapid and lasting gains from solving illegal fishing. Nat Ecol Evol 2018; 2(4): 650-8.
[http://dx.doi.org/10.1038/s41559-018-0499-1] [PMID: 29572526]

CHAPTER 3

Integrated Multi-Trophic Aquaculture Systems (IMTA): A Pathway Towards Achieving Sustainability

Hino Fernando E.[1,*], **M. Rajeshwaran**[2] and **Mohammed Meharoof**[2]

[1] *ICAR- Krishi Vigyan Kendra, TNJFU, Nagapattinam, Tamil Nadu, India*

[2] *ICAR-Central Institute of Fisheries Education, Mumbai, Maharashtra 400061, India*

Abstract: Sustainable mariculture is the practice of utilizing marine resources in an integrated method using technologies with economic and environmental sustainability without exploiting the natural resources needed for future generations. The growth of mariculture is doubling every decade, but the management and responsibility of those undertaking the activity to ensure sustainability are declining. The development of culture technology for marine organisms under controlled conditions in open ocean environments, beyond significant coastal influence, is a topic of the hour in mariculture. A new approach in mariculture that invites much-focused research on ensuring sustainability includes seed production technology, cost-effective feed, grow-out culture technology, and the social responsibility of stakeholders. Mariculture is the fastest-growing food-providing sector for the global population, with a minimal focus on environmentally sustainable practices regarding the degradation of coastal environments and the management of coastal resources. Integrated mariculture combines various farming methods, including Integrated Multitrophic Aquaculture (IMTA) and aquasilviculture, which involves combining mangrove planting with aquaculture. Modern integrated mariculture should be developed to facilitate the sustainable expansion of this sector in coastal and marine ecosystems, thereby addressing the global increase in seafood demand with a new paradigm of more efficient food production systems. Hence, mariculture as a whole must focus its growth on harnessing coastal resources more effectively by ensuring sustainability and the responsible use of marine resources.

Keywords: Fish production, IMTA, Mariculture, Marine resources, Sustainability.

[*] **Corresponding author Hino Fernando E.:** ICAR- Krishi Vigyan Kendra, TNJFU, Nagapattinam, Tamil Nadu, India; E-mail: hino@tnfu.ac.in.

G. Chelladurai, M. Paripooranaselvi & B. Makesh Kumar (Eds.)

INTRODUCTION

With a global production of 178.5 million tonnes, the fisheries sector remains a significant contributor to the global economy. It remains a major source of food, nutrition, income, and livelihoods for millions of people worldwide [1]. The contribution of total marine fisheries production in 2018 reached its highest record at approximately 65 percent (115.2 million metric tonnes), reflecting a 3.6 percent increase over the previous year, and 63.3 million metric tonnes in global inland fisheries. It is evident that capture fisheries account for around 73 percent of the total marine production and play a vital role in meeting current demand. However, global marine fisheries production has been showing a declining trend since the 1980s, with 31 percent of stocks being overfished and 58 percent fully exploited [2], which will significantly affect the livelihoods of millions of people who depend on the sector for food, nutrition, and income. Overexploitation of resources, induced by improvements in technology, juvenile fishing, bycatch, discards, and pollution, is a major challenge in the marine fisheries sector and threatens sustainability [3]. The increase in global demand for fish protein and vulnerabilities in livelihoods and the economy have driven the discovery and adoption of coastal aquaculture and mariculture techniques and technologies for candidate species to expand marine production, offering large potential for future sustainable growth [4]. Currently, mariculture and coastal aquaculture together contribute approximately 27 percent of the total marine production (30.8 million metric tonnes) and generate USD 106.5 billion [1]. Despite the development of many production technologies for marine finfish, a lion's share of 56.2 percent (17.3 million metric tonnes) of the total production is mainly dominated by mollusks, followed by finfish (7.3 million metric tonnes) and crustaceans (5.7 million metric tonnes), which underscores the immense scope for finfish mariculture.

Asian countries are dominant players in aquaculture production, and India, with its varied resources and potential, enjoys a prominent position as the second-largest producer of fish in the world, contributing approximately 6 percent of the total production [1, 5]. However, the stagnant phase of the country's marine capture fisheries sector poses a challenge in meeting the increasing demands of the fish-eating population, which has an annual per capita fish consumption of 8-10 Kg [6]. This also threatens the livelihoods and employment of 5 million fishers [7]. Mariculture can be a viable alternative for increasing seafood production, thereby meeting the country's demand and ensuring the livelihood security and income source for coastal communities. Blessed with a coastline length of 8,118 km, a continental shelf area of 0.5 million square kilometers, and a brackish water resource of 1.16 million hectares [5], the tropical country of India offers immense potential for the expansion of mariculture activities, which could become a flourishing sector shortly. However, the country is still in its early stages when compared to the global scenario. Furthermore, in the wake of the Pradhan Mantri Matsya Sampada Yojana (PMMSY) under Atma Nirbhar Bharat, with a 20 crore package to revive the blue economy [8], offers immense potential for the expansion of mariculture activities in the country.

Marine aquaculture is a specialized branch of aquaculture that involves the cultivation of economically important marine plants and animals in the sea or other natural water bodies with tidal effects, and includes onshore facilities such as brood banks, hatcheries, nursery rearing, and growing systems using seawater [9]. Mariculture plays a crucial role among coastal communities in many developing countries, contributing significantly to livelihoods, employment, and economic growth. The earliest attempts to establish saltwater fish farms were made by the Department of Fisheries, Tamil Nadu, on Krusadai Island and in the Adyar estuary [10]. Further significant interventions and contribution from several institutes like Central Marine Fisheries Research Institute (ICAR-CMFRI), Central Institute of Brackishwater Aquaculture (ICAR-CIBA), Central Salt and Marine Chemicals Research Institute (CSIR-CSMCRI), National Institute of Ocean Technology (NIOT) and the Marine Products Export Development Authority (MPEDA) have developed the mariculture of fishes, prawns, mussels, pearl oysters and sea weeds. Currently, seed production and culture technologies for a variety of finfish and shellfish species have been developed. The candidate fin fish species for mariculture includes John's snapper (*Lutjanus johnii*), Cobia (*Rachycentron canadum*), Silver pompano (*Trachinotus blochi),* Orange spotted grouper (*Epinephelus coioides*), Indian pompano (*Trachinotus mookalee*), Asian sea bass (*Lates calcarifer*), Greasy grouper (*Epinephelus tauvina*), Milk fish (*Chanos chanos*), Grey mullet (*Mugil cephalus*), and Silver moony (*Monodactylus argenteus*). The crustaceans include Indian white shrimp (*Penaeus indicus*), Green tiger shrimp (*Penaeus semisulcatus*), Mud crab (*Scylla serrata*), (*Portunus pelagicus),* Blue swimmer crab, Shovel-nosed lobster (*Thenus unimaculatus*) [11, 12]. Apart from this, the culture of seaweeds like *Gracilaria edulis, Gelidiella acerosa, Sargassum wightii, Acanthophora spicifera, Ulva lactuca*, and *Kappaphycus sp*. has developed in India. Major species of molluscs include mussels like *Perna indica* and *Perna viridis,* and clams like blood clam *(Anadara granosa)* and short neck clam (*Paphia malabarica),* Asiatic hard clam (*Meretrix meretrix) and* fertile venus *(Marcia opima), and* the edible oysters include *Crassostrea madrasensis, Crassostrea cucullata,* and *Crassostrea gryphoides* [13]. Marine pearl culture and ornamental fish culture are other potential sectors in mariculture, which could serve as a valuable source of foreign exchange. Some important species include *Pinctada fucata* and *Pinctada margaritifera,* whereas ornamental fish families include Pomacentridae *(Clownfishes and damselfishes)*, Labridae *(Wrasse)*, Scaridae *(Parrotfishes)*,

Chaetodontidae *(Butterflyfishes and Bannerfishes)*, and Acanthuridae (*Surgeon* fishes).

Fisheries and aquaculture are one of the key sectors for achieving many of the goals in the 2030 Agenda for Sustainable Development, and mariculture could aptly fit into the recognized realm of fish production, which is a sunrise sector for increasing seafood production shortly. In particular, scaling up mariculture production will help end poverty and hunger (SDG 1, 2). Gender equality can be achieved through women's empowerment in mariculture farming (SDG 5), which results in overall growth and ensures direct and indirect employment opportunities in the sector (SDG 8). Furthermore, marine resources, ecosystems, and biodiversity (SDG 14, 15) will have a significant influence by promoting sustainable mariculture development. The enhanced production and productivity in mariculture, when developed sustainably with due consideration for social and environmental dimensions, have an immense scope for achieving a blue revolution in the country.

INTEGRATED MULTI-TROPHIC AQUACULTURE SYSTEMS (IMTA): A PATHWAY TOWARDS ACHIEVING SUSTAINABILITY

Aquaculture is the fastest-growing food-producing sector in the 21st century. Aquaculture in the marine environment is an alternative solution to capture fisheries production, which could help stabilize the protein requirements of the growing global population, especially in developing countries like India. The demand for fish consumption among people is growing rapidly, and the supply side's ability to meet this demand is a significant question among fisheries scientists. The availability of land and water resources is shrinking, and the intensification of fish culture methods and feed-based culture is a potential solution for increasing fish production through aquaculture systems, particularly in marine environments, which are largely underutilized in the long term.

Mariculture production in India stands at around 28.7 million tonnes of food fish, which contributes 35.8% of the world's food fish aquaculture production. Based on the available area, the estimated mariculture production in the Indian region is 4 to 8 million tonnes compared to the current production of less than 0.01 million tonnes [14]. Mariculture is the fastest-growing food-producing sector, which is a driving force in developing the socio-economic level of local communities, as identified by the Convention on Biological Diversity (1992). The convention reiterated the possible impacts of the mariculture industry if it were taken up on a large scale. The convention highlighted the use of nutrients and antibiotics, waste disposal in mariculture activities, the transmission of disease from cultured to wild stocks, and the displacement of indigenous and local communities as key concerns. Shrimp farming, as a coastal aquaculture, has reached its maximum production limit and is losing momentum in the food production sector, as mariculture activities are increasingly being promoted by government organisations nowadays. The Twelfth Five-Year Plan of India identifies mariculture as a developing food sector that requires prompt, large-scale support. Nevertheless, the growth of the mariculture industry has invited many questions about its possible impact on the waste generated from it [15].

Most natural resources used by humans are overutilized to levels that cannot be sustained for future generations. Sustainability has become an important topic of discussion in recent times. It refers to meeting the needs of the present generation without compromising the ability of future generations to meet their own needs. It is a holistic approach that addresses ecological, social, and economic aspects, recognizing that all factors must be considered for long-term well-being. Following sustainable practices in the exploitation of marine resources could have a significant long-term impact. The United Nations Convention on the Law of the Sea (UNCLOS) primarily focuses on the conservation and sustainable use of marine resources, establishing a legal framework that contributes to nutrition and food security through their responsible exploitation. Integrated Multi-Trophic Aquaculture (IMTA) is based on the concept of utilizing excess organic and inorganic nutrients generated by intensive aquaculture systems involving species from different trophic levels. These nutrients, which would otherwise go unused, are absorbed by other organisms, thereby reducing nutrient load and preventing eutrophication.

Nutrient accumulation resulting from fish farming can be substantial and have an adverse effect on benthic communities due to smothering and increased organic load, leading to changes in sediment chemistry that cause and affect benthic biodiversity. Improving the digestibility of fish feeds and implementing computerized feed management systems are among the efforts to reduce the nutrient load surrounding fish farms. Despite technological developments, the problem of nutrient pollution associated with fish farming cannot be put to an end. A promising solution is the application of Integrated Multi-Trophic Aquaculture (IMTA) [16].

This chapter provides an in-depth examination of the history, concept, species selection, IMTA system design, and several other topics.

History and origin of the IMTA concept

The history of IMTA dates back to the emergence of aquaculture. In the year 2200-2100 B.C., the integration of fish culture with plant and vegetable production was initiated in China, as documented in the *You Hou Bin*. Moreover, there is evidence of tilapia being cultured in integrated aquaculture ponds during the New Kingdom era in Egypt, from approximately 1550-1070 B.C., as documented in tombs built during that time. IMTA was practiced at the Palace of Fontainebleau during the French Renaissance by constructing a carp pond, which is still used for fish culture. French King Henri IV had ordered that the estate be self-reliant and not depend on provisions, which were vulnerable to devastation during the 65-km trip from Paris. The complete book on Agriculture *(Nong Zheng Quan Shu)* by Xu Guangqi, in collaboration with Jesuit missionaries, was published posthumously in 1639. Irrigation and rotation of aquatic plant production, integration of livestock with fish, effect of manure application in fish pond, integrated production of mulberry trees and paddy cum fish ponds are some of the notable works done by him. John Ryther is known for his work on "integrated waste recycling marine polyculture systems" at Woods Hole Oceanographic Institution in Massachusetts. Furthermore, at the Harbor Branch Oceanographic Institute in Florida, USA, interest in IMTA was revived, which can be regarded as the grandfather of modern IMTA. After decades of technological transformations in various fish culture systems, including polyculture, integrated aquaculture, mariculture, aquariculture, ecologically engineered aquaculture, and ecological aquaculture, Dr. Chopin and Jack Taylor formulated the term "integrated multi-trophic aquaculture" in 2004 [17].

Definitions of IMTA

IMTA is similar to the concept of an integrated farming system in the vicinity, which includes species of different trophic levels within an interdependent ecosystem. This system works by utilizing uneaten feed or fecal output, which is converted and used as a fertilizer or nutrient for other organisms through mutual interaction.

IMTA is a method that combines the cultivation of fed aquaculture species, such as fish or shrimp, with organic extractive species (*e.g.*, shellfish or herbivorous fish) and inorganic extractive species (*e.g.*, seaweed) to form a balanced ecosystem. This approach promotes environmental sustainability (biomitigation), economic stability, and social acceptability by minimizing negative impacts on the surrounding environment. This method differs slightly from polyculture systems, where species are cultured within the same biological and chemical processes, often leading to ecosystem shifts. As the name suggests, multi-trophic systems involve the incorporation of species from different trophic levels within a single integrated system. IMTA represents a highly intensive culture approach that combines species from various subsystems. In IMTA, nutrient uptake and energy transfer between subsystems occur through the water medium [18].

According to Chopin (2013), IMTA is based on the principle that the solution to nitrification is not dilution, but rather extraction and conversion through diversification, in line with the law of conservation of mass formulated by Antoine Laurent de Lavoisier in 1789. The key idea is: nothing is created, nothing is lost, everything is transformed.

Integrated Multi-Trophic Aquaculture (IMTA) is a technique that involves cultivating organisms from different trophic levels together, which helps reduce effluents and increase the productivity of a given farm area. This system typically includes carnivorous finfish species cultured alongside extractive species such as macroalgae and bivalves. IMTA thus offers a means to reduce unwanted byproducts from intensive fed aquaculture, primarily Particulate Organic Matter (POM), such as fecal waste and uneaten feed pellets, and dissolved nutrients like ammonium [19].

Criteria for Species Selection in IMTA

The selection of suitable species is key to the success of any aquaculture practice. In addition, environmental sustainability should be a key criterion when utilizing natural resources for fish culture. Carnivorous fish and shrimp are typically raised using pellet feed or trash fish. Extractive organisms such as bivalves and seaweeds, which derive their nutrition from the surrounding environment, are also economically important cultured species. Since IMTA combines multiple species in the same environment, certain conditions must be followed:

Compatibility with Other Species: Species that can utilize the waste of other species as feed for growth could be an advantage, as it improves water quality and reduces pollution. Species that complement each other on different trophic levels can be selected for IMTA.

Habitat Adaptation: Species that belong to a local geographical range can be selected for culture in IMTA, as it can reduce the problem of invasiveness, which can cause harm to the environment and lead to economic losses in the culture system.

Culture, Technology, and Environmental Conditions: Species that grow on significant biomass can extract much of the excess nutrients from water through biofiltration. If it is to have fewer numbers, then species of high commercial value can be cultured; however, the bio-mitigation role is reduced in this case.

Marketability of the Species: High-market-value species should be prioritized for IMTA. Fish farmers must be able to market alternative species to enhance their economic returns. Buyers for the produce should be identified before investing in the system.

Commercialization Prospective: Select species that could create new market avenues for which policy makers and market regulators facilitate support, not to impose new regulatory restrictions for commercialization.

Sustainability: It is a prime concern when selecting the species for IMTA. The environmental performance of the environment should be improved.

- **Compatible with social and political issues**

Structure of IMTA

For the successful operation of IMTA, it is essential to properly select the site, arrange and position system components, and choose species capable of absorbing and assimilating the particulate matter and nutrients released from the fish culture system (Fig. **1**). The system should be structured or optimized so that the selected species actively recaptures the waste produced. Fish grown in cages are typically fed with formulated pellets. Uneaten feed and faecal matter settle down below the cages. Deposit feeders, such as sea urchins and sea cucumbers, take up these organic wastes. Simultaneously, suspended particles arising from these wastes are filtered out of the water by filter-feeding organisms like scallops, oysters, and mussels. Inorganic nutrients, such as nitrogen and phosphorus, are extracted by seaweeds.

They are placed away from the site in the direction of water flow so that they can utilize it properly.

Economically viable and proper density of species should be maintained throughout the culture to equalize the uptake and usage of nutrients. This system can be operated in land-based, marine, or freshwater systems, as well as in temperate or tropical waters, and may comprise multiple species [20]. Some IMTA systems included combinations of shellfish/shrimp, fish/seaweed/shellfish, fish/shrimp, and seaweed/shrimp. Filter-feeding organisms act as bioremediators, extracting nutrients, including organic and inorganic compounds. Molluscs are the most sought-after species, which have the capability of filtering macroalgae, phytoplankton, and suspended inorganic particles [21]. IMTA is a method in which farmers earn additional income by growing commercial crops. The only deviation from the usual polyculture method is the inclusion of species from different trophic levels in the same system.

Due to anthropogenic activities and excessive nutrient input, macroalgae have become efficient organisms for breaking down pollutants in the water through bioremediation. Integrated multi-trophic aquaculture is gaining importance along the Indian coast, driven by the growth of mariculture activities. A combination of commercially viable and economically important seaweed species, such as *Kappaphycus alvarezii, Gracilaria dura*, and *G. edulis*, can be cultivated with fish species like seabass, grouper, pompano, and cobia, as well as with lobsters. The red algae *Gracilaria* spp and the green algae *Ulva* spp have also been found to be efficient. IMTA in the open sea is a new start. The combined culture of compatible species of prawns and fish is of considerable importance in the context of augmenting yields in the field and the effective utilization of available ecological niches within the pond system. Cages erected within the bivalve farms, which include *Etroplus suratensis* (Pearl spot), resulted in high survival rates and growth of finfishes in the cages.

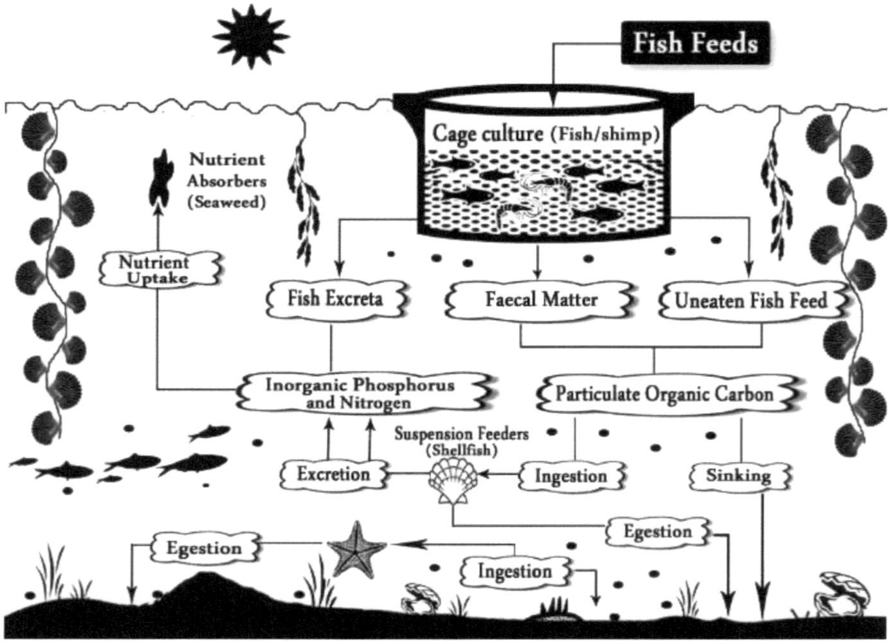

Fig. (1). Photo showing the model of IMTA.

Merits of IMTA

- IMTA can improve the local economy of the coastal people by generating employment opportunities through product processing and distribution.
- As IMTA involves the culture of various commercial species, the economic value of the output it generates can increase. Money realization happens in more

than one unit; therefore, if one species fails to achieve its level, the other species can support or compensate for the loss.

- Waste nutrients, which would otherwise be lost in conventional culture methods, are the primary source of nutrients.
- Reducing the effect of bio-mitigation, which benefits the environment by reducing eutrophication.
- Inclusion of seaweed in IMTA can prevent diseases in farmed fish due to its antibacterial activity against pathogenic bacteria.
- Since the system is operated with an organic method, there is a potential for differentiation of IMTA products through eco-labelling or organic certification.

Demerits of IMTA

- IMTA system requires a large investment as it involves technological and engineering aspects to bring out the structure.
- Aquaculture has the potential to reduce pressure on fish resources, while IMTA offers benefits to enterprises and fish farming, which compete with other users for scarce coastal resources. Additionally, stakeholder conflicts are common among local communities, leading to pollution and impacting wild fish stocks.
- IMTA, operated by different farmers, requires coordination and collaboration among production and management activities.
- Strict coastal regulations in certain countries prevent the exploitation of coastal resources for mariculture activities, especially IMTA.
- Mariculture activities in the coastal zone areas are frequently damaged and polluted by industrial effluents.
- Even though large-scale production of marine fishes poses a major challenge in IMTA, legislation is not permitted for the co-cultivation of multiple species in closeness.

Mariculture as an Alternative Livelihood Option for Resource-poor Fishers

Mariculture is one of the potential options to improve the livelihoods of coastal people and provide alternative livelihood options for fishermen, as well as ways to reduce destructive fishing practices [22]. Mariculture, as a cultural practice, involves numerous challenges that need to be addressed. Some of the major challenges are listed here [23].

CHALLENGES INVOLVED IN THE DEVELOPMENT OF MARICULTURE

Site Selection

Site selection is the major challenge in determining the commercial viability of mariculture. Identifying a location with the most favorable water quality (oxygen, temperature, light, and nutrient levels), current movements, and other infrastructure facilities are the most significant factors in mariculture. The most important aspect to be considered for selecting a suitable site for mariculture is (i) Depth and seabed characteristics, (ii) Condition of the sea. The cages, nets, and mooring system were designed to resist the storm [24].

Species Selection and Seed Availability

The two major inputs for mariculture are seed and feed [25]. A major problem for mariculture is its dependence on wild sources for seed [26].

Seed production technologies have not been scaled up, as the high cost of hatchery facilities makes the production of many high-value marine shellfish and finfish complex and expensive. Additionally, other major factors, such as live feed culture, nutritional enhancement, feeding protocols, nursery rearing, water quality maintenance, and disease management, also have a severe impact on seed production. The production of commercially important seeds through the development of commercially viable technologies is necessary for the advancement of sustainable mariculture practices. Technologies for mariculture activities are currently in the introductory phase and may take several years to become standardized in a cost-effective manner. High-value species such as Asian sea bass, for which both hatchery-produced seeds and naturally available fingerlings are accessible in various regions of India, are ideal candidates for cage stocking. Capture-Based Aquaculture (CBA) serves as an alternative to hatchery-based seed production.

As hatchery technologies have not been developed for many species, fish farmers must rely on wild seed. Capture-based aquaculture has developed in response to market demand for high-value species whose life cycles are insufficient for commercial-scale production. In the CBA, the species that can be harvested as wild juveniles include milkfish, seabass, shrimps, pomfrets, mullets, groupers, red snappers, and lobsters. The further development of mariculture is possible only by an increase in the mass production of juveniles. The supply and availability of capture-based juveniles will fulfil the demand from the world's coastal aquaculture. The species include pomfrets, mackerel, seer fish, Koth, and shrimps. A small amount is used for lobster fattening from the captured live juveniles. It is estimated that about two million mackerel juveniles and one million seer fish juveniles are landed by shore seine in April alone along the stretch of Visakhapatanam- Kalingapatnam. Only a small amount of seed is brought in live condition; it forms a very good source of CBA without affecting fishermen's livelihood and the ecosystem. It will be more profitable for the fishermen while also contributing to an increase in mariculture production. Juvenile yellowfin tuna are plentiful in and around Lakshadweep waters, which can be utilized for cage farming [24].

Feed

Availability of cost-effective and nutritionally balanced feed is another constraint for mariculture. There is no indigenous, scientifically developed marine finfish feed. The development of feed is also very complicated, and several factors need to be considered. These factors include the nutritional balance for carnivorous fish, feed conversion, and cost-effective feed. The existing feeds used for mariculture operations are not economically viable [24]. Marine fish require a high-protein and fat diet [25].

The trash fish is used as the main feed source in mariculture activities [27]. It could lead to destructive fishing practices, environmental degradation, conflicts, and impact on people's livelihoods [26]. Another problem is that the selection of feed for mariculture has a shorter storage life, seasonal fluctuations in supply, and pollution. The feed is considered an important source of disease [27]. The problem could be addressed through the development of partial or full feed alternatives. Like small-scale feed, production of these alternatives provides an opportunity for stakeholders to become engaged in mariculture practices [27] [26].

Lack of a Mariculture Policy

Mariculture is not utilized properly due to the lack of policy measures on the use of open water bodies. Coastal areas are heavily dependent on fishermen and are densely populated. The involvement of the local community is crucial for initiating mariculture activities in coastal areas. Commercializing the mariculture technology will depend on suitable policy measures [28].

Health Issues and Poor Health Management Practices

Lack of capacity in disease diagnosis: Mariculture practices deal with a variety of species, and diversified culture operations make it difficult to diagnose diseases. Moreover, there is a lack of expertise in disease diagnosis.

Unavailability of fish vaccines: On a commercial scale, there is a lack of vaccines for marine fish. The major benefits of vaccination are that vaccines should provide long-term protection and have a less significant impact on the environment.

Improper use of chemicals and antibiotics: Irresponsible use of chemicals and antibiotics leads to residue problems, rising consumer concerns, and the development of drug resistance [29].

Manpower

For the successful management of mariculture, a skilled and unskilled labor source is a major requirement. Considerable efforts, such as capacity building of human resources to meet the preferred knowledge, standards, and technical capabilities, are the immediate need for sustainable mariculture operations. In India, there is a lack of expertise in managing open-sea cage culture farms [28]. Additionally, there is a shortage of technically trained personnel to manage the multidisciplinary functions of mariculture [30].

Impact of Mariculture Activities

Mariculture activities can disturb biodiversity at the species, genetic, and ecosystem levels, resulting in undesirable impacts. The main effects include the decline of wild populations, the introduction of non-indigenous species, biological pollution, habitat degradation, genetic impacts on target species, and social effects such as human health issues and the loss of income for traditional fishermen. The waste generated from the mariculture system includes particulate nitrogen, phosphorus, organic matter, pesticides, residues of antibiotics, and hormones that enter the water column [31]. Pollution and environmental degradation have the potential to impact mariculture [23].

Suggestions for Sustainable Mariculture Practices

- Although the capital investment is low, it is necessary to provide subsidies at least during the early years to the farmers at a suitable time [32].
- Common storage facilities should be developed.
- The marketing chain was arranged within and outside the country
- Value addition should be encouraged for mariculture.
- Credit facilities are arranged by the financial institution at the lowest interest rate.
- The minimum floor price should be fixed.

- Successful mariculture practices lead to women's empowerment, increased food production, employment, and poverty alleviation. The sustainable mariculture activity will provide alternative livelihood options for fishermen and double their income [32].
- Innovations in cage culture, suitable for Indian conditions, can lead to an improvement in fish production [24].
- Increasing the public-private partnership in mariculture will help address the financial imbalance. Indigenous micro feed production is encouraged to fulfill the demand for mariculture feed [25].
- The adverse effect of mariculture activity is reduced by the proper environmental assessment, proper feeding protocol, better genetic resource management, effective site selection, better effluent, and waste control measures, setting up of hatcheries for seed collection, reducing the collection of wild seed, and increasing the positive effect of mariculture to reduce the stress on capture fisheries [31].
- For an efficient mariculture industry, the following recommendations are suggested: development of integrated farming systems, reduction of fish meal and fish oil inputs in feed, expansion of farming for lower trophic level fishes, and promotion of environmentally friendly mariculture practices for resource management [33].
- The lack of seed for cultural practices is addressed through the development of small-scale hatcheries [23, 26, 34]. The advantages of Small-scale hatcheries include relatively low capital costs, ease of management, flexibility, and the ability to use a variety of marine fish species, quick economic returns, accessible technology, and a focus on larval rearing and fingerling production [34].
- Sustainable development of mariculture requires skill-building in health management and market diversification, as well as financial support [28].

FUTURE DIRECTION

- R & D in sustainable mariculture practices.
- Development of a fish disease surveillance system.
- Broodstock management and seed quality improvement for cultured species.
- Standardized seed production technology and technology upgradation are needed for the development of sustainable mariculture. The seed production technologies should be suitable for open-cage farming.
- The development of mariculture can be achieved by establishing seed banks in the appropriate location. Seed banks should be developed for commercial species. Wild seed could be used for the development of a seed bank.
- The massive open-sea cage culture should be encouraged by central and state government support.
- Integrated multi-trophic aquaculture could be a viable solution to reduce the

environmental impact and increase economic viability.

- Mariculture technologies should be popularized among multiple stakeholders.
- There is a good demand for agar and carrageenan on the global level. This demand could be met by large-scale seaweed farming in the Palk Bay region.
- India is one of the richest sources of ornamental fishery resources, especially in the Gulf of Mannar. Seed production technologies should be developed for marine ornamental fish species.

CONCLUSION

Small-scale fisheries would be the worst affected if onshore marine resources are degraded. Increased pressure on natural fish stocks and a decline in marine fish production over the years have had a severe impact on Small-scale fishers and poor coastal communities. It leads to destructive fishing practices and insecurity of livelihoods. Therefore, there is a need to promote sustainable income-generating activities beyond fishing. It would be better to have a mariculture-based activity as an alternative livelihood option for the fishers. This would encourage them to have a better understanding of the environment, and a community fisheries management model should be applied to achieve better resource conservation. Mariculture is one approach to addressing this problem. IMTA offers a suitable alternative for enhancing marine fish production through sustainable methods, particularly for coastal fishers. A people-centered approach is one way to achieve sustainability in mariculture activities. It could also be incorporated into integrated coastal management plans to reduce resource use conflicts and promote sustainability [23].

REFERENCES

[1] Food and Agriculture Organization. The State of World Fisheries and Aquaculture. Sustainability in action. Rome 2020.

[2] Food and Agriculture Organization. The State of World Fisheries and Aquaculture. Sustainability in action. Rome 2016.

[3] Pauly D, Christensen V, Guénette S, *et al.* Towards sustainability in world fisheries. Nature 2002; 418(6898): 689-95.
[http://dx.doi.org/10.1038/nature01017] [PMID: 12167876]

[4] Oyinlola MA, Reygondeau G, Wabnitz CCC, Troell M, Cheung WWL. Global estimation of areas with suitable environmental conditions for mariculture species. PLoS One 2018; 13(1): e0191086.
[http://dx.doi.org/10.1371/journal.pone.0191086] [PMID: 29357374]

[5] Department of Fisheries. Ministry of Fisheries, Animal Husbandry & Dairying, Government of India, Handbook of Fisheries Statistics 2018.

[6] Shyam SS. Fish consumption pattern in India: Paradigm shifts and Paradox of export trade (Fish consumption pattern in India, exports-Overview). Food and Beverage News 2016; pp. 25-8.

[7] Pillai NGK. Marine fisheries and Mariculture in India. Marine Fisheries and Mariculture in India 2011; pp. 1-326.

[8] Meharoof, M., Gul, S., & Qureshi, N.W. Indian Seafood Trade and Covid-19: Anticipated Impacts and Economics. Food and scientific reports 2020; 1(8): 54-58. 19.

[9] Gopalakrishnan A, Kirubagaran R, John G, *et al.* Draft National Mariculture Policy 2019 (NMP2019). CMFRI Marine Fisheries Policy Series No. 17. Prepared by P. Boby Ignatius, Committee constituted by the National Fisheries Development Board (NFDB), Ministry of Fisheries, Animal Husbandry &a Dairying, Government of India. Kochi, India: Central Marine Fisheries Research Institute; 2019.

[10] Selvaraj GSD. Mariculture in India, its potentialities and practical applications. Seafood Export J 1973; 5: 29-37.

[11] Central Institute of Brackishwater Aquaculture (CIBA) - Annual Report 2018-19.

[12] Central Marine Fisheries Research Institute(CMFRI). Annual Report 2018-19, 304 p.

[13] Ignatius, B., Rajesh, N. and Imelda, J. Technological advancement in Mariculture in India for production enhancement and sustainability, 2020.

[14] National Fisheries Development Board (NFDB). National Mariculture Policy 2019 [Internet]. 2019 [cited 2020 Dec 15]. Available from: http://nfdb.gov.in/PDF/Revised%20draft%20-%20NMP-2019.pdf

[15] Naveen Namboothri, C.M. Muralidharan and Aarthi Sridhar. Mariculture and Food Production: Sustaining the promise. POSITION PAPER, 2012. (Cited: 12 December, 2020). Available from: https://www.dakshin.org/wp-content/uploads/2017/06/Mariculture_Final_0910.pdf

[16] Wang X, Olsen LM, Reitan KI, Olsen Y. Discharge of nutrient wastes from salmon farms: environmental effects, and potential for integrated multi-trophic aquaculture. Aquacult Environ Interact 2012; 2(3): 267-83.
[http://dx.doi.org/10.3354/aei00044]

[17] Chopin T. Integrated multi-trophic aquaculture Ancient, Adaptable Concept Focuses On Ecological Integration. Glob Aquacult Advocate 2013; 16-8. Available from: https://www2.unb.ca/chopinlab/articles/files/Chopin%202013%20GAA%20IMTA.pdf

[18] Barrington, K., Chopin, T. and Robinson, S. Integrated multi-trophic aquaculture (IMTA) in marine temperature waters. In: Integrated mariculture: a global review (ed. Soto, D.). FAO Fisheries and Aquaculture Technical Paper 2009; 529: 7-46.

[19] Alexander KA, Freeman S, Potts T. Navigating uncertain waters: European public perceptions of integrated multi trophic aquaculture (IMTA). Environ Sci Policy 2016; 61: 230-7. b
[http://dx.doi.org/10.1016/j.envsci.2016.04.020]

[20] Neori A, Chopin T, Troell M, *et al.* Integrated aquaculture: rationale, evolution and state of the art emphasizing seaweed biofiltration in modern mariculture. Aquaculture 2004; 231(1-4): 361-91.
[http://dx.doi.org/10.1016/j.aquaculture.2003.11.015]

[21] Marinho-Soriano E, Azevedo CAA, Trigueiro TG, Pereira DC, Carneiro MAA, Camara MR. Bioremediation of aquaculture wastewater using macroalgae and Artemia. Int Biodeterior Biodegradation 2011; 65(1): 253-7.
[http://dx.doi.org/10.1016/j.ibiod.2010.10.001]

[22] Haylor, G., Briggs, M.R.P., Pet-Soede, L., et al.Improving coastal livelihoods throughsustainable aquaculture practices. In NACA, ed. A report to the collaborative APEC grouper research and development network 2003 (FWG/01/2001). pp. 1-35. Bangkok, NACA.

[23] Gonzales, E.M., MaungSoe, U.K., Mukherjee, R., et al. Regional review on livelihood opportunities related to mariculture development. In A. Lovatelli, M.J. Phillips, J.R. Arthur and K. Yamamoto (eds). FAO/ NACA Regional Workshop on the Future of Mariculture: a Regional Approach for Responsible Development in the Asia-Pacific Region. Guangzhou, China, 7-11 March 2006. FAO Fisheries Proceedings. No. 11. Rome, FAO. 2008. pp. 65-94.

[24] Rao, G.S. Overview on mariculture and the opportunities and challenges of cage culture in India,

2009.

[25] Vijayagopal, P. Nutrition, Feeds and Feed Technology for Mariculture in India, 2015.

[26] Suspita A. The current status of mariculture in Indonesia and livelihood opportunities for coastal communities The future of mariculture: a regional approach for responsible development of marine farming in the Asia-Pacific region. Bangkok: STREAM 2006.

[27] Sim, S.Y., Rimmer, M.A., Toledo, J.D., *et al.* A practical guide to feeds and feed management for culturedgroupers. Bangkok 2005b; NACA.

[28] Bhat, B.V. and Vinod, P.N. Development of seafarming in India – an export perspective. In A. Lovatelli, M.J. Phillips, J.R. Arthur and K. Yamamoto (eds). FAO/ NACA Regional Workshop on the Future of Mariculture: a Regional Approach for Responsible Development in the Asia-Pacific Region. Guangzhou, China, 7-11 March 2006. FAO Fisheries Proceedings. No. 11. Rome, FAO. 2008. pp. 301-306

[29] Tan, Z., Komar, C. and Enright, W.J. Health issues in mariculture in the Asia- Pacific region: an industry perspective. In A. Lovatelli, M.J. Phillips, J.R. Arthur and K. Yamamoto (eds). FAO/NACA Regional Workshop on the Future of Mariculture: a Regional Approach for Responsible Development in the Asia-Pacific Region. Guangzhou, China, 7-11 March 2006. FAO Fisheries Proceedings. No. 11. Rome, FAO.2008; pp. 321-322.

[30] James PSBR. Manpower requirement in mariculture. Proceedings of the symposium on management of coastal aquaculture and oceanic resources of Andamans 1987; 55-9.

[31] Joshi KK, Thobias PA, Varsha MS. Mariculture and biodiversity. 2015.

[32] Mohan Joseph M. Mussel and oyster farming: a small step in Indian mariculture but a giant leap in coastal rural livelihood. Fishing Chimes 2003; 23(1): 39-41.

[33] Naylor RL, Goldburg RJ, Primavera JH, *et al.* Effect of aquaculture on world fish supplies. Nature 2000; 405(6790): 1017-24.
[http://dx.doi.org/10.1038/35016500] [PMID: 10890435]

[34] Sim SY, Rimmer MA, Toledo JD, *et al.* A guide to small-scale marine finfish hatchery technology. Bangkok: NACA 2005. a

CHAPTER 4

Mariculture Pollution and Its Impacts on Biodiversity

M. I. Delighta Mano Joyce[1,*], P. J. Joslin[2], C. Stella Packiam[3] and M. Vijayalakshmi[4]

[1] *Department of Zoology, Sadakathullah Appa College (Autonomous), Tirunelveli, Tamil Nadu, India*

[2] *Department of Zoology, St. Mary's College (Autonomous), Thoothukudi, Tamil Nadu, India*

[3] *Department of Chemistry, A.P.C. Mahalaxmi College for Women, Thoothukudi, Tamil Nadu, India*

[4] *Department of Zoology, A.P.C.Mahalaxmi College for Women, Thoothukudi, Tamil Nadu, India*

Abstract: Mariculture is an activity primarily focused on producing aquatic food, and many people living along the coast depend on it for their livelihood and income. Due to population growth and rising demand for seafood products, it has rapidly expanded. Maricultural activities discharge particulate organic matter, nitrogen, phosphorus, antibiotics, and hormones into the marine ecosystem. Large areas of mangrove and coastal regions have been converted into shrimp and fish ponds. It also has unintended predation impacts on non-target species. These have the potential to alter, destroy, or disrupt habitat, trophic structures, disease transmission, and genetic capabilities. As a result, mariculture affects biodiversity, which will be reflected in the supply of ecosystem services. Mariculture development must be sustained to promise economic and environmental benefits.

Keywords: Mariculture, Biodiversity, Habitat modification, Pollution.

INTRODUCTION

Mariculture is the practice of cultivating marine organisms in their natural habitats for commercial purposes. Mariculture contributes to the production of protein-rich foods, providing a source of income for millions of coastal communities. It is one of the fastest-growing food-producing sectors, offering high-quality food and being more efficient than many other food-production systems around the world.

* **Corresponding author M. I. Delighta Mano Joyce:** Department of Zoology, Sadakathullah Appa College (Autonomous), Tirunelveli, Tamil Nadu, India; E-mail: midelighta@gmail.com

G. Chelladurai, M. Paripooranaselvi & B. Makesh Kumar (Eds.)

The nutritional importance of fish has been acknowledged since the dawn of humanity. Humans can get plenty of protein from fresh fish, which contains all of the important amino acids, minerals, and vitamins in the amounts that humans require. It also has a pleasant taste and flavor and is simple to digest.

Mariculture is dominated by unfed aquaculture of aquatic plants, bivalves, and mollusks, as well as fed aquaculture of marine finfish and shrimp. The potential interaction and conflict between mariculture activities and marine ecosystems have been a long-standing concern, with aquaculture-derived effluents being one of the most significant factors contributing to negative environmental consequences [1]. Mariculture systems generate significant volumes of effluent that may have negative environmental implications due to the recurring input of feeds, fertilizers, and chemicals required to sustain the health and growth of the culture [2, 3]. The effluent contains large amounts of organic wastes, nutrients, and chemicals, which are discharged into receiving waters, then carried through the water column and deposited on the bottom, affecting the nearby area and beyond [4, 5]. Hypoxia, eutrophication, heavy metal contamination, and habitat destruction are just a few of the problems that they can create [6]. As a result, to manage the discharge of mariculture effluents and their repercussions, adequate and rigorous regulatory measures are required [7, 8]. Habitat degradation, the collapse of wild populations, the introduction of non-indigenous species, biological pollution, and genetic effects on target species are among the most significant consequences. Additionally, social repercussions include health difficulties and the loss of employment and income for traditional fishermen.

It is well acknowledged that all forms of mariculture harm biodiversity at the species, genetic, and ecological levels. There are several viable options for avoiding the negative consequences of mariculture on biodiversity. These include effective site selection, proper environmental assessment, proper feeding protocols, improved effluent and waste control measures, enhanced genetic resource management, the establishment of seed hatcheries, reduced wild seed collection, and the enhancement of the positive effects of mariculture to reduce pressure on capture fisheries [9].

Mariculture Pollution

Due to advances in technology, the development of formulated feeds, and a deeper biological understanding of farmed species, as well as improvements in water quality within closed farm systems, rising demand for seafood products, site development, and increased government attention, mariculture has experienced rapid growth over the past two decades [10 - 12]. As a result, there has been some debate about the social and environmental consequences of mariculture [13, 14].

The degree of environmental impact, like most farming activities, is determined by factors such as the size of the farm, the species being cultivated, stock density, type of feed, the site's hydrography, and husbandry practices [15].

Cage Culture's Waste

Waste products from cage aquaculture operations in freshwater are discharged directly into aquatic habitats. These wastes can take the form of particulate debris (*e.g.*, feces and uneaten feed) that fall to the benthic habitat beneath the cage arrays, or dissolved wastes that are discharged directly into the water column from fish feed, feces, and metabolic excretions. Several factors may influence the amount of waste produced, including fish size, water temperature, and aquaculture practices (*e.g.*, feed composition, diet, and feeding methods). Phosphorus is typically the most limiting nutrient for algal growth in freshwater ecosystems; therefore, its release into the environment is of particular concern. Excessive phosphorus input into lakes can stimulate algal blooms, leading to harmful or nuisance growth, shoreline fouling from attached algae, and reduced oxygen levels in deeper waters.

Finfish farming may necessitate a high-protein food supply. Due to inefficient feeding regimes and poor digestibility of formulated diets, a large amount of fishmeal is wasted, resulting in low feed conservation ratios. Overfeeding primarily results in increased debris deposition on the bottom in coastal fish farms, whereas excess food goes to waste in hatcheries and land-based farms, potentially impacting the adjacent catchment and local coastal ecology. This effect is highly dependent on the settling velocity of the waste feed, as well as the current velocity and depth [16, 17].

Farm Escapees

Escapees from aquaculture operations have a negative impact on local ecosystems by causing hybridization and loss of genetic diversity in native populations, as well as increasing negative interactions such as predation, competition, disease transfer, and habitat alteration. Farmed mollusks become prominent predators and competitors in molluscan farming, potentially spreading infections and parasites [18].

Disease, Parasite Transfer, and Genetic Pollution

One of the most serious issues with mariculture is the risk of disease and parasite transmission. Aquaculture stocks are frequently selectively bred to enhance resistance to diseases and parasites, improve growth rates, and increase product quality. However, if these cultured organisms escape, they may introduce pathogens and parasites (*e.g.*, lice) into wild populations. This can lead to disease outbreaks, as wild species may lack immunity to these new infections. Additionally, genetic pollution from escaped stocks may reduce the ability of wild populations to adapt to changing natural environments. Cultivated non-indigenous species may also carry or be resistant to illnesses that could be devastating to native species. Such escapes can lead to increased predation, competition, and habitat alteration, which can severely affect local ecosystems [19, 20].

Habitat Alteration

Shrimp cultivation contributes only a minor amount to the deterioration of mangrove habitats. Mangrove forests produce nutrient-dense matrices that support a diverse range of wildlife, primarily juvenile fish and crabs. By processing organic material and filtering sediments, they act as buffering systems, reducing coastal erosion and improving water quality for *in situ* organisms [21, 22].

Construction Materials

Heavy metals, plastic compounds, and by-products are all released into the marine environment by some construction materials. Farmers are often unaware of the presence of these compounds. However, we are becoming increasingly conscious of them every day. Fatty acid salts, chromates, cadmium sulfide, antioxidants, UV stabilizers, organophosphates, fungicides, and disinfectants are all found in plastics. The majority of these composites are poisonous to aquatic life; however, due to their low solubility and slower degradation rate, their toxicity is reduced. Antifouling paints on docks, vessel bottoms, and nets contain an effective and inexpensive chemical called tributyltin, which is the active ingredient in most antifouling paints [23]. Tributyltin pollution impairs oyster shell production, kills mussel larvae, induces imposex and endocrine disruption in mollusks, and contaminates sediment, seawater, fish, and other bivalve species [24].

ENVIRONMENTAL IMPACTS

Water Quality

Solid wastes, such as uneaten pellet feed, feces, fish scales, mucus, and soluble wastes like ammonium and urea, are commonly found in aquaculture wastes. Nitrogen, phosphate groups, and suspended particles are the primary components that affect water quality among these pollutants [25].

Pollution Caused by Organic Matter

Fish feces, uneaten feed, and deceased animals are the primary sources of organic contaminants. The ingredients in fish feed are linked to their feces. Cage systems limit the speed of the water flow, causing organic compounds to accumulate in the surrounding environment. The decomposition of these organic pollutants causes water quality to deteriorate and increases the oxygen requirements of the water. Turbidity, the deposit of organic waste in the sediment, the formation of anoxic zones along the seafloor, the accumulation of hazardous compounds, and the spread of illnesses are all factors to consider. The formation of gases such as hydrogen sulfide, methane, and carbon dioxide in the sediment decreases the growth rate of living organisms in the area, making them more susceptible to disease [26]. The direct degradation of organic waste from aquaculture activities can reduce dissolved oxygen levels in the water.

Pollution by Chemicals

A wide range of chemicals is employed in aquaculture activities for disinfection and disease outbreak therapy [27]. The toxic effect of a chemical molecule, pathogen sensitivity to antibiotics and therapies, and the chemical's active duration in the environment all influence its environmental impact. Chemicals are released directly into the water column and sediment in the majority of chemical treatment scenarios in mariculture. The most popular fish species in cage systems are Atlantic salmon, rainbow trout, sea bass, and sea bream. The majority of the chemicals used in their breeding cause pollution.

Chemical Impacts

Pollution from organic and inorganic nutrients, as well as from anthropogenic sources such as hazardous chemicals, medicines, and metals, has a chemical influence. These factors can alter sediment composition, lead to eutrophication, and induce hypoxic or anoxic conditions. The emission of inorganic waste is the most commonly addressed and demonstrated chemical and overall form of damage generated by suspended mariculture.

Antibiotics

In aquaculture, bioactive compounds such as antibiotics and insecticides are commonly used to combat diseases, parasites, and control the spread of pathogenic bacteria [28 - 31]. When antibiotics are used excessively over time, this raises concerns about the probable consequences on biological diversity and the selection of antibiotic-resistant microorganisms [29]. Furthermore, while the use of a single medicine may have little impact, when combined with other chemical substances, it can have a considerable negative effect [32]. Antibiotics administered as medicated feed are predicted to end up in the marine environment near farming locations at a rate of 70 to 80 percent [33]. Parasiticides, which pollute the water column and bottom as uneaten medicated feed, feces, or in a soluble form in urine, pose another environmental danger [34]. Furthermore, parasiticides used in aquaculture to manage ectoparasite infestations may not be specific to target organisms, posing a risk to non-target species such as small crustaceans in the vicinity of the farm site [29]. Even though the use of antibiotics and organophosphates in aquaculture has decreased, synthetic parathion, colorants, preservatives, anti-parasitic agents, and other chemicals that pollute the ocean continue to be used [35, 36]. The use of chemicals has an impact on the marine environment and poses a risk to farmers [37]. The correct use of chemicals determines the success or failure of aquaculture, the duration of their application, and the conditions under which they are used to combat infectious illnesses and parasites.

Impacts on Sediment

Uneaten feed and faeces gather on the seafloor after feeding fish in a cage. Excessive accumulation of organic material on the seafloor leads to an increase in the number of microbes, which reduces the amount of oxygen available to other benthic species (crustaceans, fish, *etc.*). Because of the accumulation of uneaten feed, faeces, toxins, and antibiotics on the seafloor, most benthic creatures that reside around salmon farms vanished, while just a few resistant bacteria proliferated [38]. Antibiotics used to combat bacterial fish illness outbreaks are typically discharged into the water column [39]. These antibiotics build up in the sediment or on animals in the environment [40]. In aquaculture, the most commonly used antibiotics are oxytetracycline, florfenicol, and sulfonamides. The oxytetracycline in feed cannot be completely digested. After metabolism, some of the active ingredients are expelled in feces and subsequently released into the water [37].

Impacts of Mariculture on Sediment

Research on the effects of pesticides, antibiotics, and synthetic paints on human health has received less attention than studies on the effects of these chemicals on the marine environment. Uneaten feed and faeces gather on the sea floor after feeding fish in a cage. Excessive buildup of organic material on the seafloor leads to an increase in the number of microbes, which reduces the amount of oxygen available to other benthic species [41].

Impacts on Microbial Communities

Heterotrophic bacteria play a significant part in biodegradation in the marine ecosystem [42]. Bacteria are extremely sensitive to environmental changes; therefore, anthropogenic nutrient input can directly affect them [43, 44]. One of the most significant influences on bacterial communities is the use of chemical drugs. Antibiotic-added feed and medicines released directly into the marine environment exert selection pressure on bacterial populations and fish infections, leading to the development of antibiotic resistance genes [45, 46]. Antibiotics have a significant impact on the microbial organisms that live in the sediment. Antibiotic use alters bacterial biomass and the balance of different bacterial types. Bacteria that live in sediment control the nitrogen, phosphate, and sulfate cycles, as well as the first step in the food chain. Antibiotic contamination lowers sulfate and nitrate production rates.

Impacts on Biodiversity

Mariculture has the potential to alter, degrade, or destroy habitats, as well as disturb trophic systems, decrease natural seed stocks, spread diseases, and reduce genetic diversity. Coastal mangroves have been converted into shrimp ponds, nutrient loading has impacted enclosed or semi-enclosed waters, and benthic ecosystems have been affected by bivalve bottom culture activities, as well as sedimentation [47].

Antifouling products are used in aquaculture to prevent biofouling by different epibiotic organisms such as marine algae, barnacles, bivalves, bryozoans, and others on submerged structures (cage nets and supporting infrastructure). Biofouling in marine finfish aquaculture can weaken cage structures and have severe consequences for farmed fish, reducing water flow through cages and thereby decreasing oxygen availability or increasing the risk of illness. Increased copper levels have been linked to antifouling chemicals, which are hazardous to a variety of marine invertebrates, particularly mollusks. Antifouling and copper-based chemicals also influence the immune response of exposed species around aquaculture areas [34, 48]. There is limited information on the amounts of disinfectants and specific compounds, and there have been few studies on the effects of disinfectants used on land-based facilities, nets, platforms, boats, ponds, and other surfaces [49].

In marine aquaculture, anesthetics are used to alleviate stress during handling and increase the efficiency of processes [50]. However, due to occasional and restricted use and almost no discharge into the environment when sound management techniques are followed, the use of these chemicals does not pose a substantial threat to the environment [29]. Biological contamination can occur as a result of the unintentional release of fish during operation, cage breakage, or attacks by wild animals. It may also occur during spawning, when farmed fish (*e.g.*, Gadus morhua) are maintained in open cages until they reach sexual maturity, allowing fertilized eggs to float into the surrounding environment [51]. Although unintentional escapes (particularly large-scale ones) are uncommon, when compared to the abundance of their wild conspecifics, such as wild Atlantic salmon, the number of individuals fleeing can be significant [52]. Ecological interactions have negative consequences for wild populations, including potential genetic implications from interbreeding and competition for food and habitat [53]. Selective breeding to enhance growth performance accounts for 80-83% of European aquaculture production [54]. In the wild, farmed fish may pass on these genetic traits, resulting in shorter lifespans, increased susceptibility to illness, and reduced individual fitness, all of which may contribute to a lower survival rate in future wild populations.

Drastic Reduction in O_2 level

In open waters, culture systems discharge their nutrient-rich waste directly into the water, potentially increasing the trophic status. Semi-enclosed environments, such as sheltered bays, are particularly vulnerable. This could result in phytoplankton blooms, including hazardous species, and the breakdown of these blooms could significantly lower oxygen levels. Algal blooms can also shade seafloor vegetation, which provides a nursery habitat and a sanctuary for finfish and benthic invertebrates [55].

The majority of scallop and oyster faeces will settle under the beds. During storms, the material can be carried up into water columns, obstructing the bivalves' gills and causing mass mortality. This sediment overlay can also alter the composition of benthic communities, favouring pollution-tolerant species, resulting in a noticeable impact on biodiversity [56, 57].

Chemicals are also utilized as disinfectants and antifouling agents. Antifoulants like TBT are banned for use in aquaculture in affluent countries, but they are still employed in some developing countries, where they have a negative impact on biodiversity. Sandnes and Ervik (1999) reported that, except for copper, which increased from 119 tonnes in 1991 to 174 tonnes in 1995, salmon production in Norway has increased, while antifoulant usage has decreased [58].

Parasites in cultivated stock are a nuisance for both aquaculturists and other organisms in the environment. Acquisition from a fish farm, for example, is one possible source of Parvicapsula infection in migrating Pacific salmon in British Columbia. The parasite is thought to be associated with significant changes in salmon migratory behavior, which result in high pre-spawning mortality and may be responsible for population-wide genetic variation loss. Small fish are harvested for conversion into fish meal, leaving fewer predatory fish, such as cod, and other marine predators, including seabirds and seals, in the food web [59].

Bivalve culture depletes the marine food web of nutrients. However, it only harms biodiversity if the carbon and nitrogen removed from the water column become excessive, leaving less for other herbivores and phytoplankton, which in turn affects the growth and reproduction of zooplankton and other herbivorous marine animals.

Permanent, widespread bivalve production may cause eutrophication by altering the coastal food web [60, 61]. When habitat loss or alteration affects the living conditions of other species, it becomes a biodiversity issue. Seed harvesting using damaging gear from ecosystems, such as lagoon bottom habitats, results in habitat destruction. Mariculture requires a substantial amount of space, and the sheer number of acres of water occupied can impact the migratory paths and feeding patterns of non-target species. Some people argue that salmon farms, for example, obstruct the free flow of wild migrating salmon while also feeding killer whales. Non-target animals may be stressed by underwater explosives and other noise devices meant to dissuade predators [62].

Mangroves and tidal marshes, which serve as nursery grounds for wild shrimp and fish populations, are disappearing, and mangrove and marsh grass waste is entering coastal food webs in smaller amounts [63].

CONCLUSION

Mariculture can produce high-quality products for human consumption and benefit the economy by creating jobs in various fields. It is essential for our country, and with proper management, it can also benefit the environment. Maricultural activities discharge particulate organic matter, nitrogen, phosphorus, antibiotics, and hormones, among other substances, into the marine ecosystem. These discharges have the potential to alter, destroy, or disrupt habitats, trophic structures, disease transmission, and genetic diversity, thereby affecting biodiversity. Mariculture development must be managed sustainably to ensure both economic and environmental benefits.

REFERENCE

[1] Kluger LC, Filgueira R Thinking outside the box: embracing social complexity in aquaculture carrying capacity estimations. ICES J Mar Sci. 2020;77(6):2083–2091.

[2] Bambaranda BVASM, Tsusaka TW, Chirapart A, *et al.* Capacity of *Caulerpa lentillifera* in the removal of fish culture effluent in a recirculating aquaculture system. Processes (Basel) 2019; 7(7): 440.
[http://dx.doi.org/10.3390/pr7070440]

[3] Joesting, HM; Blaylock R; Biber P; Ray A, The use of marine aquaculture solid waste for nursery production of the salt marsh plants *Spartina alterniflora and Juncus roemerianus*. Aquac 2016. Rep. 3, 108–114.

[4] Bannister RJ, Johnsen IA, Hansen PK, Kutti T, Asplin L. Bannister RJ, Johnsen IA, Hansen PK, Kutti T, Asplin L, Near- and far-field dispersal modeling of organic waste from Atlantic salmon aquaculture in fjord systems. ICES J. 2016 Mar. Sci. 73, 2408–2419

[5] Pusceddu A, Fraschetti S, Mirto S, Holmer M, Danovaro R. Effects of intensive mariculture on sediment biochemistry. Ecol Appl 2007; 17(5): 1366-78.
[PMID: 17708214]

[6] Carballeira C, Cebro A, Villares R, Carballeira A. Assessing changes in the toxicity of effluents from intensive marine fish farms over time by using a battery of bioassays 2018.

[7] Primavera JH. Overcoming the impacts of aquaculture on the coastal zone. Ocean Coast Manage 2006; 49: 531-45.

[8] Crab R, Avnimelech Y, Defoirdt T, Bossier P, Verstraete W. Nitrogen removal techniques in aquaculture for sustainable production. Aquaculture 2007; 270: 1-14.

[9] Joshi KK, Antony TP, Varsha MS. Mariculture and Biodiversity. Winter School on Technological Advances in Mariculture for Production Enhancement and Sustainability. GSJ 2021; 9(1): 281-9.

[10] DeVoe MR. Aquaculture and the marine environment: policy and management issues and opportunities in the United States. Bull Natl Res Inst Aquacult Supp 1994; 1: 111-23.

[11] Read P, Fernandes T. Management of environmental impacts of marine aquaculture in Europe. Aquaculture 2003; 226(1–4): 139-63.

[12] Ross A. Leaping in the Dark: A Review of the Environmental Impacts of Marine Salmon Farming in Scotland and Proposals for Change. Perth, Scotland: Scottish Environment Link 1997.

[13] Ervik A, Hansen PK, Aure J, *et al.* Regulating the local environmental impact of intensive marine fish farming I. The concept of the MOM system (Modelling-Ongrowing fish farms-Monitoring). Aquaculture 1997; 158(1–2): 85-94.

[14] Jennings S, Kaiser MJ, Reynolds JD. Marine Fisheries Ecology. Blackwell, Victoria 2001

[15] Wu R. S. S. The environmental impact of marine fish culture: Towards a sustainable future. Marine Pollution Bulletin. 1995, 31 (4–12): 159–166.

[16] Forrest B, Keeley N, Gillespie P, et al. Review of the ecological effects of marine finfish aquaculture: final report. Prepared for Ministry of Fisheries. Cawthron Report No. 1285. Nelson, New Zealand: Cawthron Institute; 2007.

[17] Black KD. Mariculture, Environmental, Economic and Social Impacts of.Encyclopedia of Ocean Sciences. Academic Press 2001; pp. 1578-84.

[18] Naylor RL, Williams SL, Strong DR. Ecology. Aquaculture--a gateway for exotic species. Science 2001; 294(5547): 1655-6.
[PMID: 11721035]

[19] Rosenberg AA. Aquaculture: the price of lice. Nature 2008; 451(7174): 23-4.
[PMID: 18172486]

[20] Fent K, Weston AA., Caminada D. Ecotoxicology of human pharmaceuticals. Aquatic toxicology 2006 (Amsterdam, Netherlands) 76(2):122-59.

[21] Kaiser MJ, Attrill MJ, Jennings S, *et al.* Marine Ecology: Processes, Systems, and Impacts. New York: Oxford University Press 2005.

[22] Trujillo AP, Thurman HV. Essentials of Oceanography. 9th ed., New Jersey: Pearson Prentice Hall 2008.

[23] Burton NF, Day MJ, Bull AT. Distribution of bacterial plasmids in clean and polluted sites in a South Wales river. Appl Environ Microbiol. 1982;44(5):1026–1029.

[24] Hawkey PM, Jones AM. The changing epidemiology of resistance. J Antimicrob Chemother 2009; 64 (Suppl. 1): i3-i10.
[PMID: 19675017]

[25] Carroll ML, Cochrane S, Fieleri R, Velvin R, White P. Organic enrichment of sediments from salmon farming in Norway: environmental factors, management practices, and monitoring techniques. Aquaculture 2003; 226: 165-80.

[26] Shumway SE. Toxic algae: A serious threat to shellfish aquaculture. World Aquaculture Society 1989; 20: 65-74.

[27] Braithwaite RA, Carrascosa CMC, McEvoy LA. Biofouling of salmon cage netting and the efficacy of a typical copper-based antifoulant. Aquaculture 2007; 262: 219-26.

[28] McCracken A, Fidgeon S, O'Brien JJ, Anderson D. An investigation of antibiotic and drug residues in fish. J Appl Bacteriol 1976; 40(1): 61-6.
[PMID: 1270364]

[29] Burridge L, Weis JS, Cabello F, Pizarro J, Bostick K. Chemical use in salmon aquaculture: A review of current practices and possible environmental effects. Aquaculture 2010; 306: 7-23.

[30] Porchas MM, Cordova LRM. World aquaculture: environmental impacts and troubleshooting alternatives. 2012; Volume: Article ID 389623.

[31] Rico A, Phu TM, Satapornvanit K, Min J, Shahabuddin AM, *et al.* Use of veterinary medicines, feed additives and probiotics in four major internationally traded aquaculture species farmed in Asia. Aquacuture 2013; 412: 231-43.

[32] Fent K, Weston AA, Caminada D. Ecotoxicology of human pharmaceuticals. Aquat Toxicol. 2006;76(2):122–159.

[33] Ferreira I., Horst KVD., Vos WW., *et al.*, Environmental correlates of physical activity in youth – a review and update 2006.

[34] Tornero V, Hanke G. Chemical contaminants entering the marine environment from sea-based sources: A review with a focus on European seas 2016.

[35] Austin B, Austin DA. Bacterial Fish Pathogens The disease of Farmed and Wild Fish. 5th ed., Springer Science 1998.

[36] Kluger LC, Filgueira R. Thinking outside the box: embracing social complexity in aquaculture carrying capacity estimations. ICES J Mar Sci. 2020;77(6):2083–2091.

[37] Hakanson L, Wallin M. Use of ecometric analysis to establish load diagrams for nutrients in coastal areas.Copenhagen Marine Aquaculture and Environment Nordic Council of Ministers. 1991; pp. 9-23.

[38] Komoike Y, Matsuoka M. Exposure to tributyltin induces endoplasmic reticulum stress and the unfolded protein response in zebrafish. Aquat Toxicol 2013; 142-143(142-143): 221-9.
[PMID: 24055755]

[39] Crozier WW. Escaped farmed salmon, *Salmo salar L.*, in the Glenarm River, Northern Ireland: genetic status of the wild population 7 years on. Fish Manag Ecol 2000; 7(5): 437-46.

[40] Alzieu C, Heral M. Ecotoxicological effects of organotin compounds on oyster culture.Ecotoxicological testing for the marine environment. Belgium: State University 1984; pp. 187-95.

[41] Morton AB, Symonds HK. Displacement of Orcinus orca (L.) by high amplitude sound in British Columbia, Canada. ICES J Mar Sci. 2002;59:71–80.

[42] Beaumont AR, Budd MD.High mortality of the larvae of the common mussel at low concentrations of

tributyltin 1984. Marine Pollution Bulletin 15: 402-405.

[43] Morcillo Y, Porte C. Evidence of endocrine disruption in clams--Ruditapes decussata--transplanted to a tributyltin-polluted environment. Environ Pollut 2000; 107(1): 47-52. [PMID: 15093007]

[44] van Wezel AP, van Vlaardingen P. Environmental risk limits for antifouling substances. Aquat Toxicol 2004; 66(4): 427-44. [PMID: 15168950]

[45] Jaque MB, Caceres WC, Muniesa M. Transfer of antibiotic-resistance genes via phage-related mobile elements. Plasmid. 2015;79:1–7.

[46] Hada HS, Sizemore RK. Incidence of plasmids in marine *Vibrio spp.* isolated from an oil field in the northwestern Gulf of Mexico. Appl Environ Microbiol 1981; 41(1): 199-202. [PMID: 16345685]

[47] CBD Technical Series. Secretariat of the Convention on Biological Diversity Solutions for sustainable mariculture - avoiding the adverse effects of mariculture on biological diversity 2004, No. 12

[48] Fitridge I, Dempster T, Guenther J, Nys RD. The impact and control of biofouling in marine aquaculture: a review. Aquacult Res. 2012;43(7):649–669.

[49] Buschmann AH, Riquelme VA, González MCH, Varela D, Jiménez JE, *et al.* A review of the impacts of salmonid farming on marine coastal ecosystems in the southeast Pacific. *ICES J.* Mark Sci 2006; 63: 1338-45.

[50] Sneddon LU. Clinical anaesthesia and analgesia in fish. J Exot Pet Med 2012; 21: 32-43.

[51] Jensen Ø, Dempster T, Thorstad EB, Uglem I, Fredheim A. Escapes of fishes from Norwegian sea-cage aquaculture: causes, consequences and prevention. Aquacult Environ Interact 2010; 1(1): 71-83.

[52] Thorstad EB, Heggberget TG, Økland F. Migratory behaviour of adult wild and escaped farmed Atlantic salmon, Salmo salar L., before, during and after spawning in a Norwegian river. Aquacult Res 29(6): 419-28.

[53] Gómez ID, Lopz AP, Sempere JTB, Jerez PS. Assessing the influence of gilthead sea bream escapees in landings of Mediterranean fisheries through a scale-based methodology Fish Manag Ecol 24: 67-72.2017;

[54] Janssen K, Chavanne H, Berentsen P, Komen H. Impact of selective breeding on European aquaculture. Aquaculture 2017; 472: 8-16.

[55] Staniford D. Yes: it's true, pollutants pollute. New Scientist. 2002;175(2356):30.

[56] Kapuscinski A R, Nega T, Hallerman E M. Adaptive biosafety assessment and management regimes for aquatic genetically modified organisms in the environment 1999.

[57] Mattsson J, Linden O. Benthic macrofauna succession under mussels, Mytilusedulis L (Bivalvia), cultured on hanging longlines Sarsia 68:97-102. Cited in Goldberg 1983.

[58] Sandnes K, Ervik A. Industrial marine fish farming.Sustainable aquaculture: food for the future?. Rotterdam: A. A. Balkema 1999; pp. 97-107.

[59] Pauly D, Froese R, Liu LY, Tyedmers P. 2001.Down with fisheries, up with aquaculture? Implications of global trends in the mean trophic levels of fish.

[60] Horsted S. J., T. G., Nielsen B., Riemann J., Steen P., and Bjornsen P.K. Regulation of zooplankton by suspension-feeding bivalves and fish in estuarine enclosures 1988. Mar. Ecol. Prog. Ser. 48:217-224

[61] Tenore KR, Boyer LF, Cal RM, *et al.* Coastal upwelling in the Rias Bajas, NW Spain: contrasting the benthic regimes of the Rias de Arosa and de Muros. J Mar Res 1982; 40: 701-72.

[62] Morton AB, Symonds HK. Displacement of *Orcinus orca* (L.) by high amplitude sound in British Columbia, Canada. ICES J Mar Sci 2002; 59: 71-80.

[63] Naylor RL, Goldburg RJ, Primavera JH, *et al.* Effect of aquaculture on world fish supplies. Nature 2000, 405:1017-1024. Cited in Smith Oceanography & Fisheries Open Access Journal 2017; 4(1): 25-32.

CHAPTER 5

Disease Management for Sustainable Mariculture

Anisha Valsalam[1,*]

[1] *ICAR – Central Institute of Fisheries Education, Department of Aquatic Environment Management, Mumbai, Maharashtra, India*

Abstract: In the fisheries sector, mariculture serves as a growing backbone for the nutritional animal producing industry. Despite the progression in the farming practices for production maximisation, loss occurs through diseases by a variety of pathogens. Hence, disease management is an important criterion that must be implemented at the proper time for the prevention of diseases and also to protect cultured organisms from pathogens. Aquatic health management requires adequate knowledge to overcome the practical problems encountered in the mariculture sector. The important pathogens are categorised into major groups such as viruses, bacteria, fungi, and parasites. The viral diseases in the mariculture environment predominate with the iridoviral disease, lymphocystis disease, nervous necrosis virus disease, and rhabdoviral disease. The major bacterial diseases encountered include vibriosis, winter ulcer, photobacteriosis, and furunculosis. One of the leading causes of other infections in the saline aquaculture practices is parasites, especially in cage cultures. Disease clearance requires some practical prophylactic and therapeutic actions to eradicate them. Disease treatments using chemicals cause various problems to the fish by settling as residues in the body, polluting the environment or harms the other organisms in the surrounding environment. To avoid this and to effectively eradicate the problematic organisms, some new environmentally friendly measures can be taken to control the pathogens from the mariculture environment. Therefore, the drawbacks of using chemicals for treatments can be reduced, and better results can be expected to ensure a sustainable mariculture environment.

Keywords: Bacteria, Disease management, Parasites, Pathogen, Therapeutics.

INTRODUCTION

Globally, the disease occurrence in the mariculture environment is continuously increasing with an immediate downward effect on the wellness of the culture organisms [1]. The causative agents of the diseases are broadly categorised as bacteria, viruses, fungi, and parasites (Fig. **1**). Though fungal disease outbreaks

* **Corresponding author Anisha Valsalam:** ICAR – Central Institute of Fisheries Education, Department of Aquatic Environment Management, Mumbai, Maharashtra, India E-mail: anisha.aahpa901@cife.edu.in

G. Chelladurai, M. Paripooranaselvi & B. Makesh Kumar (Eds.)

are not well documented in the mariculture environments, the sporadic occurrences of significant oomycete outbreaks are unavoidable in the natural settings, which require additional research insight in the field to gain a better understanding [2]. Bacterial pathogens are one of the significant causes of production loss in the culture systems along with viral disease outbreaks. The acute viral outbreaks may result in total crop loss by causing sudden mortality of the culture organisms. Hence adequate management and preventive strategies must be employed to prevent and to overcome the disease outbreak situation. Parasitic infestations are a significant problem encountered in all the marine cage cultured areas [3]. Parasites have found the culture environment as a suitable place of attachment [4], and the severity of the infestation compels the aquaculturist to use chemical remedies which need to be chosen carefully after consideration, as this may affect the surrounding environment and the associated organisms [5]. Proper discarding of the used products, other chemicals, and wastes must be ensured to avoid future issues with the environment and ecosystem thereof.

The condition of disease is regarded as the outcome of a complex interaction between the host, pathogen, and the environment as these components interplay among themselves [6]. The factors of the pathogen contributing to disease severity include the ability to persist, zoonotic nature, host specificity, host range, virulence, attachment, colonisation, presence of toxins, and variance of infectivity of the genotype. Various environmental factors contribute to the predisposition of diseases such as water quality parameters, climate, and meteorology. The host-associated factors of disease occurrence include life stage, genetic factors, immune impairment, incorrect diet, and stress [7]. Production costs are found to be high due to disease outbreaks. Disease outbreaks are challenging to be avoided entirely in mariculture but can be inhibited, and incidence can be minimised. There are two broad classifications of diseases based on the cause, such as infectious diseases caused by pathogenic organisms and non-infectious diseases. Various stress factors depress the fish's immune system, which in turn encourages the numbers of opportunistic pathogens [6].

Stress can have apparent effects on the health of fishes. Stress can appear in many forms such as dietary stress due to inappropriate diet and feeding schemes, environmental stress, and physical stress attendant to handling, or confinement. The environmental stress factors can be categorised as physical, chemical, procedural, and biological [7]. The unfavourable environmental conditions which cause severe stress in fish include overcrowding, sub-lethal levels of toxic materials, temperature fluctuations, low dissolved oxygen, salinity fluctuations, high concentrations of obnoxious gases like hydrogen sulfide and ammonia, excessive or rough handling, *etc* [7]. Crowding is the most important predisposing factor for disease as it creates plenty of communal pressure among the competing fish species for dissolved oxygen, food, and space. Massive crowding facilitates the horizontal transmission of contagious diseases by various means. These types of pressures are inescapable in concentrated fish farming. Long-lasting experience of stress or even very brief stressful situations can lower specific cellular and humoral aspects of the immune system, thereby lowering resistance to pathogens [7]. Once stress is experienced, a flow of neuroendocrine actions occurs, leading to the rise of the steroid stress hormone, cortisol. The cortisol along with another stress hormone such as catecholamines can reduce the performance of the immune system which is carried out by reducing the number and distribution of circulating leukocytes, depressing the macrophage activity, and reducing the number of immunoglobulin-producing cells [7].

BACTERIAL DISEASES

Bacterial diseases in the marine environment are caused by various pathogens [8], such as *Renibacterium salmoninarum, Lactococcus, Streptococcus* [9], *Piscirickettsia salmonis, Pasteurella piscida, Yersinia ruckeri, Staphylococcus epidermis, Nocardia asteroids, N. campuchi,* and *Clostridium* sp [10, 11]. The most significant bacterial diseases, like vibriosis [12], winter ulcer, photobacteriosis [13], furunculosis, marine flexibacteriosis, pseudomonadiasis, edwardsiellosis, and mycobacteriosis [14] are described briefly.

Vibriosis

Vibrio spp. are the most important bacterial pathogens [15], ubiquitously distributed in the marine and estuarine culture systems causing 'salt-water furunculosis' with seasonal dynamics in population [14]. There are numerous species which are associated with pathogenicities such as *V. harveyi* [16], *V. anguillarum, V. spendidus, V. penaecida, V. alginolyticus* [17], *V. parahaemolyticus,* and *V. vulnificus* [12, 18]. The Vibrio infected fish show clinical symptoms [19], such as pale gills, dark skin, slow growth, necrosis of skin and appendages, haemorrhages at the base of fins, skin ulcers, exophthalmia, corneal opacity, internal organ liquefaction, body malformation, enteritis, blindness, muscle opacity, and splenomegaly [10].

Winter Ulcer

Characteristics of the disease include skin ulcers confined to scaled body surfaces and often petechial bleeding in internal organs [10]. Although mortality is limited, the market price of the affected fish is reduced [18]. The disease is caused by a psychrotrophic bacterial species referred to as *Moritella viscosa.* The bacterium has the thread-forming ability with adherent colonies in culture media [14].

Photobacteriosis

Photobacterium damselae is also a significant pathogen [13] causing a disease called photobacteriosis [10]. It was first isolated from damselfish (*Chromis punctipinnis*) and has spread to various species such as turbot (*Psetta maxima*), eel (*Anguilla reinhardtii*), rainbow trout (*Oncorhynchus mykiss*), ovate pompano (*Trachinotus ovatus*), red-banded seabream (*Pagrus auriga*), sea bream (*Sparus aurata*), sea bass (*Dicentrarchus labrax*), yellowtail (*Seriola quinqueradiata*), white seabream (*Diplodus sargus*), and meagre (*Argyrosomus regius*) [20]. The bacterial genus comprises two subspecies, *i.e. P. damselae* subsp. *damselae* and *P. damselae* subsp. *piscicida* in which the former is recognised as an opportunistic pathogen, while *P. damselae* subsp. *Piscicida* [21] causes severe septicaemia in marine fishes [15, 18]. The bacteria are ubiquitously distributed and are reported to have widespread antibiotic resistance. The "atypical" strain of the bacteria was found to be associated with ulcerative disease in marine fishes [14, 20].

Aeromonas Infections

Aeromonas hydrophila are found to infect warm water fishes worldwide and are considered as a significant economic problem, particularly in China and India over the past years. In coastal premises, an outbreak dominated by *A. hydrophila* was reported [8]. The clinical signs of the disease include redfin, haemorrhagic septicemia, dermal ulceration, ocular ulcerations, tail or fin rot, erythrodermatitis, and scale protrusion [10]. In the acute form of the disease, fatal septicemia may occur so quickly that the fish die before they develop deep gross signs of disease [18]. The ventral portion of the fish may become swollen as a result of oedema, and the scales may stand up. *Aeromonas salmonicida* was also found to be capable of surviving in the marine environment, causing furunculosis [14].

Marine Flexibacteriosis

The causative agent of the disease in marine fish is *Tenacibaculum maritimum*. Other names representing the disease include eroded mouth syndrome, gliding bacterial diseases of sea fish, and black patch necrosis [10]. The primary target organs of the disease are skin and gills, with the clinical signs manifested as ulcers, reddening, eroded skin, yellow mucous, frayed fins, haemorrhagic mouth, lesions, and tail rot [18]. Similar filamentous bacteria causing disease include *Leucothrix mucor, Thiothrix* sp., *Flexibacter* sp., *Flavobacterium,* and *Cytophaga* sp. The infection results in lethargy, slow growth, discolouration of gills, empty gut, and increased mortality [14].

Pseudomonadiasis

The most significant pathogens under the genus *Pseudomonas* for the cultured fishes include *P. chlororaphis, P. anguilliseptica, P. fluorescens, P. putida,* and *P. plecoglossicida* [18]. The bacteria *P. anguilliseptica* was described in 1972 as the causative agent of sekiten-bio or red spot disease of the eel in Japan. The pathogen was also isolated from other fishes, such as black seabream, ayu, salmonids, wild herring, turbot, and black spot sea bream [10]. The prominent clinical manifestations include abdominal distension and petechia in the skin and internal organs [22]. Fin rot is caused by *Pseudomonas fluorescens, P. putida,* and *P. putrefaciens.* The other typical signs include anorexia, inactivity, skin discolouration, emaciation, septicaemia, ulceration, and haemorrhages [14].

Edwardsiellosis

The disease is caused by the pathogen *Edwardsiella tarda* and results in multi-organ infection with hypoxia, swollen abdomen, body lesions, petechial haemorrhages, and skin depigmentation [10, 18]. The bacterium is found to be of zoonotic nature and results in human infection. However, the bacterium *E. piscicida* is found to be more pathogenic to fish than *E. tarda* [23].

Mycobacteriosis

The sub-acute to chronic wasting disease is also referred to as fish tuberculosis and is known to affect freshwater and salt-water species [22]. A large number of Mycobacterium species associated with granulomas in fish populations have been described, such as *M. marinum, M. simiae, M. fortuitum, M. chelonae, M. smegmatis, M. abscessus, M. poriferae, M. neonarum, M. scrofulaceum*, and *M. triplex-like* [14, 18].

VIRAL DISEASES

The primary viral diseases of the fishes include iridoviral disease, Lymphocystis disease [24], Nervous Necrosis Virus (NNV) disease, Infectious Pancreatic Necrosis (IPN) disease, Viral Hemorrhagic Septicemia (VHS) [25], Infectious Hematopoietic Necrosis (IHN), Epizootic Hematopoietic Necrosis (EHN), Spring Viremia of Carp (SVC), *Oncorhynchus masou* Virus Disease (OMVD), and Infectious Salmon Anemia (ISA) [26].

Viral Fish Diseases

Fish Iridoviral disease (IVD), caused by the *Iridovirus* primarily affects the gills, spleen, and other internal organs. The manifested clinical signs are lethargy, pale discolouration of the body, and cessation of feeding. Lymphocystis disease (LCD) or black spot disease is one of the viral diseases affecting the cultured yellowtail [18, 24]. Lymphocystis cells in the skin or fins of the fishes are found to be surrounded by abnormal melanophores or black spots. The disease mainly occurs in the summer months and primarily affects lymphocytes and epidermal cells. The typical clinical signs of the disease include cauliflower-like lesions on various body surfaces like the mouth, fin, tail, grey colouration, and aggression. The swimming capacity of the fish is affected [24]. Viral Nervous Necrosis Disease (VNND) is caused by *Nodavirus* affecting the nervous system [15], and the typical signs of the disease include cork-screwing, whirling, darting, cessation of feeding, belly-up motion, swim bladder hyperinflation, and pale grey body pigmentation [27]. Infectious Pancreatic Necrosis (IPN) disease is caused by *Birnavirus* affecting the organs such as the pancreas and the liver. The signs of the diseased fish include lying on the bottom, swimming in a cork-screw motion, whitish faecal casts, swollen belly, darkened body, pale gills, and exophthalmia [26].

Viral Hemorrhagic Septicemia (VHS) is caused by *Novirhabdovirus* affecting the vascular system and the kidney of the fishes. The perceptible clinical marks such as pastel gills, bleeding around eyes, murky body, exophthalmia, pinpoint haemorrhages, and ulceration, characterise the disease [25]. Infectious Hematopoietic Necrosis (IHN) is caused by a *Rhabdovirus* known to affect the kidney, spleen, and liver. The affected fishes exhibit dark skin, lethargy, cessation of feeding, hyperactivity, exophthalmia, white faecal discharge, and fin-base bleeding. Epizootic Hematopoietic Necrosis (EHN) is caused by a *Ranavirus*, which is known to affect the liver, kidney, spleen, and other parenchymal tissues. Distended abdomen, darkened skin colour, petechial haemorrhages at the fin base, and gills are the typical clinical signs of the disease [26]. Spring Viremia of Carp (SVC) is caused by a *Rhabdovirus,* affecting the significant organs of the fish, such as the liver, kidney, spleen, gills, and brain with the clinical signs including exophthalmia, pale gills, haemorrhages on the skin, bases of fins, and vent, dropsy and protruding anus with trailing mucoid faecal casts [28]. *Oncorhynchus masou* Virus Disease (OMVD) is caused by a *Herpesvirus* and affects the skin, gill, liver, kidney, and spleen. The clinical signs are exhaustion, reduced feeding, dark body, skin ulcers, and petechiae [18]. Tumour occurrence can be seen on the jaws, fins, operculum, cornea, and body surface. Infectious Salmon Anemia (ISA) is caused by an *Orthomyxovirus* affecting the endothelial cells in the heart, liver, spleen, and kidney. The clinical signs of the disease include pale gills, exophthalmia, distended abdomen, blood in the anterior eye chamber, scale pocket oedema, and skin haemorrhages [26].

Viral Crustacean Diseases

The predominant viral infections in the commercial shellfish farming in the coastal provinces include *White Spot Syndrome Virus (WSSV)*, *Taura Syndrome Virus (TSV)*, *Yellow Head Virus (YHV)*, *Hepatopancreatic Parvovirus (HPV)* [29], *Monodon Baculovirus (MBV)*, *Baculovirus Penaei (BP)* [30], *Infectious myonecrosis virus (IMNV), Infectious Hypodermal Haematopoietic Necrosis Virus* (IHHNV) [26], and the newly emerged *Decapod Iridescent Virus-1 (DIV-1).*

White Spot Syndrome Virus (WSSV) is the most threatening disease-causing agent in intensive shrimp culture practice. The disease is severe as the aetiological agent is highly virulent with a vast geographic and broad host range. High mortality can be noted within four to seven days of disease outbreaks with up to 100% mortality [29]. *Taura syndrome virus (TSV)* in *P. vannamei* culture causes mortality up to 80%, and the clinical feature is the presence of dotted, band like black lesions along with the abdominal outer shell [26]. *Yellow Head Virus (YHV)* is characterised by a light yellow tinge of the cephalothorax area. Generally, the affected shrimp are pale. Shrimp infected may die within a few hours of developing the clinical symptoms [30]. The complete crop loss may occur within 3-5 days. *Monodon Baculovirus (MBV)* is a widespread pathogen and is not highly virulent. The clinical manifestations are lethargy and inadequate feed intake; however, the histopathological findings reveal a circular occlusion body which is the typical diagnostic feature of the disease [29]. It causes occasional severe mortalities with up to 90% in the post-larval stage and over 70% in juvenile stages [26].

Hepatopancreatic Parvovirus (HPV) infection results in non-specific gross signs such as reduced preening activities, atrophy of the hepatopancreas, anorexia, and low growth rate. The mortality associated with this virus may range from 50% to 100% [29]. *Baculovirus penaei (BP)* was first reported in 1974 in the pink shrimp, *Penaeus duorarum* and this virus affects primarily the hepatopancreas and midgut epithelial cells [26]. The usual route of infection is *via* faecal contamination of spawned eggs from the adult spawners that are infected. *Infectious Hypodermal Haematopoietic Necrosis Virus (IHHNV)* was first reported in *P. stylirostris* and *P. vannamei* and later in *P. monodon. IHHNV* was found to infect the tissues of mesodermal and ectodermal origin [30]. Thus, the organs infected include gills, lymphoid organs, connective tissues, hypodermis, antennal gland, and nerve cord ganglia. The endodermal derived tissues are not affected by the virus. *Infectious myonecrosis virus (IMNV)* causes marked oedema in muscle fibres, fluid accumulation in the tissues, lymphoid organ spheroid formation, haemocyte infiltration, and fibrosis. The hypertrophy caused by Lymphoid Organ Spheroids (LOSs) is a consistent wound in shrimp. The ectopic LOSs are found in antennal gland tubules, haemocoelom in the gills, heart, and ventral nerve cord. *IMNV* belongs to the family Totiviridae [26].

The recent *Decapod Iridescent Virus-1 (DIV-1)* or *Shrimp haemocyte iridescent virus (SHIV)* was first reported in 2017, in Zhejiang Province, China. The disease caused high mortality in farmed *L. vannamei* and was found to have a broad host range from freshwater to marine. Based on the sequence analysis of major capsid protein (MCP) and ATPase, DIV-1 was categorised into the genus *Xiairidovirus* in the Iridoviridae family [31]. The virus infects tissues of ectodermal and mesodermal origin, including the gills, ventral nerve cord, ganglia, cuticular epithelium, connective tissues, haematopoietic tissues, antennal gland, and the lymphoid organ explicitly. The symptoms exhibited by diseased *P. vannamei* include empty stomach and guts, loss of colour or reddish body, and soft-shell. The affected shrimp lose swimming ability, and the life stages infected include post larvae to sub-adult [31].

Viral Molluscan Diseases

Viral diseases of marine molluscs reported in mariculture and the wild environment include *Papovavirus, Abalone Viral Ganglioneuritis (AVG),* and *Oyster herpes virus, Gill Necrosis Virus (GNV), Haemocyte Infection Virus (HIV), Oyster Velar Virus Disease (OVVD), Picornavirus, Birnavirus,* and *Reovirus* [30].

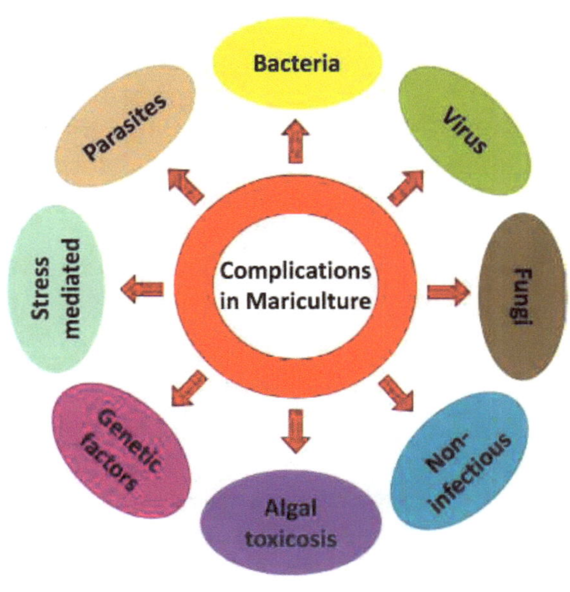

Fig. (1). Depiction of various disease complications in mariculture system.

PARASITIC INFECTIONS

The parasitic organisms are grouped into protozoans, ciliates, flagellates, myxosporidians, isopods, leeches, helminths, and copepods [32]. The protozoans are found to be adhering to the body parts such as gills, internal organs, fins, cephalothorax, pereiopods, and other appendices [5]. Protozoa may induce brown gill disease in crustaceans leading to anorexia and reduced growth. The presence of protozoans may increase the susceptibility to other infections by opportunistic pathogens. Protozoa reported in shellfish include *Zoothamnium, Epistylis, Vorticella*, *Anophrys, Acineta* sp, *Lagenophry*s, and *Ephelota.* The gregarians are found in crustaceans such as shrimp. Ciliates such as *Cryptocaryon irritans, Tricodina* spp [33, 34], *Brooklynella* spp, *Paranophrys* spp, and *Parauronema* sp, and parasitic flagellates such as *Amyloodinium ocellatum, Leptomonas* sp. are reported in marine fishes. Isopods causing parasitism are reported such as *Rhexanella sp.*, *Nerocila sp.,* and *Nerocila phaiopleura*. Leech parasites under Hirudinea reported include *Zeylanicobdella* and *Arugamensis*. *Lepeophtheirus sp* [4] affects the maricultured fish species and is a serious threat [35]. Bloody fluke aporocotylid infection is reported in marine fish [36, 37].

Myxosporidiosis

Myxosporidian diseases are caused by various species of myxosporidians [15, 33], including *Sphaerospora* spp., *Kudoa pericardialis,* and *Kudoa amamiensis* [38]. Pericardial kudoasis is caused by *Kudoa pericardialis* and is characterised by the formation of a white substance in the pericardial cavity [39]. These are the trophozoites of *Kudoa pericardialis* [5]. The parasite produces little harmful effect on the fish and does not affect the selling price because the site of accumulation is limited to the pericardial cavity [40]. Muscular kudoasis is caused by *Kudoa amamiensis,* which thrives in the muscle, producing white spherical or oval cysts [24].

Helminthous Diseases

Capsalid monogeneans affecting the skin are *Benedenia* spp., and *Neobenedenia* spp. Diplectanic monogeneans affecting the gill include *Diplectenum* spp [24]. Dactylogyrids include *Haliotrema* spp. and *Dactylogyrus* spp. Microcotylids monogenean includes *Heterobothrium* spp., and *Heteraxine* [5]. The monogenean skin fluke *Benedenia* is a harmful parasite found on cultured yellowtail. The parasite attaches tightly to the body surface and affects mainly the epithelial tissues of the host [40]. The affected fish becomes emaciated with reduced appetite and rubs the affected part owing to itchiness [15]. As a result, the skin injury provides a portal of entry for bacteria that cause vibriosis. *Heteraxine heterocerca,* the parasite, has two small anterior suckers and a large posterior sucker with clamps to adhere to the gill and suck the blood of the host [24].

Philometrosis

Philometrosis is caused by a Nematode, *Philometroides seriolae,* which bores a hole in the trunk muscles, and the larvae come under the serous membrane of the liver. The parasite has a minimal detrimental effect on the fish, but the selling value of infected fish is reduced [24]. The parasite, while in the muscle, forms abscess-like material inside a pouch. The transitional host of the parasite is thought to be a copepod [24].

Bladder Worm Disease

This disease is caused by the abdominal parasitism of plerocercus larvae of *Callotetrarhynchus nipponica* (Cestoda). The parasites are about two to three cm long and covered with the connective tissue pocket of the host. The high parasitism of the larva causes physiological disturbances in the fish and retards its growth [24].

Caligiosis

The parasitic disease caused by the copepods is called caligiosis. Copepods causing disease include *Caligus* spp [38]. (commonly called sea lice), *Lernaeocera branchialis* [41], *Pseudocaligus* spp., and *Lernanthropus latis* [24]. Caligiosis is a gill disease caused by *Caligus spinousus* (Crustacea). The carapace acts as an adhesive disc. The lunules and the second antennae also serve as adhering organs [42].

FUNGAL ENCOUNTERS

Around 500 fungal species have been described in marine and estuarine environments. In crustaceans, mostly the larval stages are affected, and the common causative agents are *Lagenidium callinectes, Serolpidium* spp., *Fusarium* spp., *Lagenidium marina*, *Sirolpidium parasitica*, *Fusarium solani,* and *F. moniliformae. Fusarium* spp. is one of the opportunistic fungal pathogens that may lead to high mortality, up to 90%. The common *microsporidian* disease is cotton shrimp disease [43]. Enterocytozoan Hepatopenaei (EHP) results in retarded growth in cultured shrimp. Infected animals also exhibit White Faecal Syndrome (WSF) and severe necrosis of the hepatopancreatic tissue.

In marine and brackish water fishes, the reported fungal diseases *include epizootic ulcerative syndrome, microsporidiosis, and ichthyophonosis* [24]. *Epizootic Ulcerative Syndrome (EUS)* is caused by the fungus *Aphanomyces invadens,* which affects the skin and internal organs. Ulcers with irregular margins and creamy to red colouration, cloudy eyes, slimy patches, and discolouration of the body are the typical clinical signs of the disease [43]. The disease is further characterised by extensive granulomatous myositis, myonecrosis, vacuolation, necrotic tissue debris, granulomatous inflammation, and debris of free hyphae, inflammatory cells in skeletal muscle, and deep focal ulcers that expose the musculature.

Ichthyophonosis

Ichthyophonosis incidences in fishes are caused by a fungus, *Ichthyophonus hoferi,* which affects all the interior organs, including the heart, liver, gills, nervous system, and spleen. The affected organs swell and show white spots. In the gills, haemorrhage with marked anaemia can be noticed [24]. The fish affected by the fungi eventually die [44].

Microsporidian Disease

The disease is characterised by an uneven fish body, low growth, and rapid weight loss. The mortality is reported to be low, and fishes heal naturally, but growth is still retarded [24]. The disease is called microsporidiosis and affects the trunk muscles of the fish, which show the formation of white lumps of trophozoites packed with large numbers of tiny microsporidian spores. The frequency of outbreaks is short, and its infectiousness is fragile [43]. Microsporidians are reported in vast numbers in fishes, for example, *Glugea* spp. and *Pleistophora* spp [24].

DISEASES IN HOLOTHUROIDEA

The following viral diseases, such as *Acute Peristome Oedema Disease (APED),* and *Stomach Atrophy Syndrome (SAS)* have been identified in sea cucumbers. The bacterial diseases, such as *Skin Ulceration Disease (SUD)* [45], *Stomach Ulceration Symptom (SUS), Viscera Rejection Syndrome (VRS), Rotting Edges Symptom (RES),* and *Off-Plate Syndrome (OPS)* have been reported in sea cucumbers. The protozoan parasitic infection of sea cucumber is caused by *Diplodina gonadipertha.* The fungal disease is caused by two varieties, such as branched and straight hyphae fungi. The copepods predate on the sea cucumbers, and the crucial copepods are from the genus *Microsetella* [30].

DISEASES IN LOBSTERS AND CRABS

Panulirus argus virus (PaV1) is reported in 2004, which affects the juvenile lobster *P. argus*. The clinical signs and pathology include milky haemolymph, lethargy, discoloured, and fouled carapace. Rickettsia-like bacteria cause the bacterial disease called the *Milky Haemolymph Disease of Spiny Lobsters (MHD-SL)* [46]. Another bacterial disease, *Gaffkaemia or red-tailed disease* affects the American lobster *Homarus americanus*. *Epizootic Shell Disease (ESD)* affecting the American lobster, *Homarus americanus,* is caused by the bacterium *Aquimarina homaria* [30].

The *Sighs Disease (SD)* in crab is a viral infection caused by a roni-like virus that infects the connective tissues. The virus is named *Eriocheir sinensis ronivirus (EsRNV)*. The primarily affected host is the Chinese mitten crab (*Eriocheir sinensis)* [30].

NON-INFECTIOUS DISEASES

There is also information on loss or disease occurrence in various aquaculture systems, which are non-infectious and happen due to various factors, such as nutritional deficiency, stress [7], environmental conditions, algal blooms [47], food intake regulations, neoplasms, and related disorders, chemically induced alterations, developmental disorders, genetic disorders, overcrowding, disordered metabolism, immunological disorders, effects of pesticides and pollutants, hydro-mineral imbalances, excess dissolved gases, and depraved water quality parameters. A better understanding of the mechanisms through which diet influences immune coordination is necessary to appreciate the many complex connections between diet and predisposition to infectious diseases [48]. Diet arrangement and physical features of the diet may alter the microorganisms in the digestive tract and the integrity of the intestinal epithelium. Inadequate diet and the unsuitable balance of nutrients within the diet can have a significant effect on immune functions [49].

Toxicity Related to Low Water Quality

Temperature-induced stress increases the disease susceptibility of fish and thereby decreases the resistance nature [7]. High alkaline waters corrode the fins, gills, and result in frays [49]. With a dissolved oxygen deficient system, the resistance mechanism of fish is no longer maintained at an adequate level [49]. In water, the noxious effect of ammonia depends directly on the pH value. The higher the pH of water, the greater is the ammonia toxicity and vice versa [49]. The H_2S toxicity is directly proportional to the water temperature, and H_2S concentration in fish ponds should be maintained below 0.002 ppm [49].

Algal Toxicosis

Recognition of toxic circumstances often goes unrecorded since such severe algal blooms frequently end in oxygen exhaustion, which is then held responsible for the resultant fish death [47]. Fish subjected to such conditions may display erratic swimming, appear tired, become convulsive, and die. Smaller fish species typically die first, but subsequently, entire populations may be killed in the surroundings [47].

DISEASE PREVENTION AND MANAGEMENT STRATEGIES

Successful fish health management begins with the prevention of disease rather than treatment. The aquatic organisms exhibit some behavioural modifications like gasping at the surface, reduced feeding, lethargy, and rubbing against objects. They also show some physical signs, like the presence of sores, ragged fins, and abnormal body conformation [50]. Prevention of disease outbreaks can be taken care of by incorporating proactive compounds in the feed that stimulate the host's immune system such as immunostimulants, neutraceuticals [51] probiotics [52], and prebiotics. Prevention of outbreaks can also be supported by constant health monitoring, stringent record-keeping, and maintaining hygiene [53]. The entry of wild shrimps, crabs, and fish into the ponds should be prevented. When an infectious disease outbreak occurs, the contaminated water should be disinfected before discharge [54]. Physical, chemical, and biological precautionary measures can be taken on the farm. Physical measures can be aimed at preventing the intrusion of disease-carrying vectors, water treatment, quarantine [55], increased aeration, controlled temperature, improved feeding regimes, removal of sludge and organic matter, and treatment of wastewater. Chemicals can be used to treat constituents before they arrive at the culture system; ionisation, chlorination, and iodine can be used for the treatment of incoming water and other potential components such as clothing, nets, tools, baskets, and footwear. Effective chemical measures include control of salinity, pH, and reduction of ammonia, nitrite, and the use of antibiotics if required. Biological precautionary measures consist of the use of probiotics to create helpful microbial populations under the culture conditions [53].

Overall System Maintenance

A variety of ecological and best management strategies can be adopted for the control of viral and other significant preventable diseases in aquaculture (Fig. **2**). Good husbandry practice, including nutrition, water, and soil quality management, can be implemented. Control strategies and regular monitoring of excess organic matter and can also be carried out [54]. The feed should be nutritionally balanced, and feeding can be adjusted to the required quantity by proper observation. It is essential to use clean, fresh feed [53]. Proper storage and handling of feed can reduce food-borne diseases such as fungal mycotoxins. For immediate reference, daily records must be available. While shipping fish for diagnosis, it is best to send the dying fish showing clinical signs to the diagnostic laboratory [50]. Cross-contamination of equipment or water from one system to another can be avoided. Restricting the entry of persons and apparatus can reduce the chance of disease transmission from one facility to another [53].

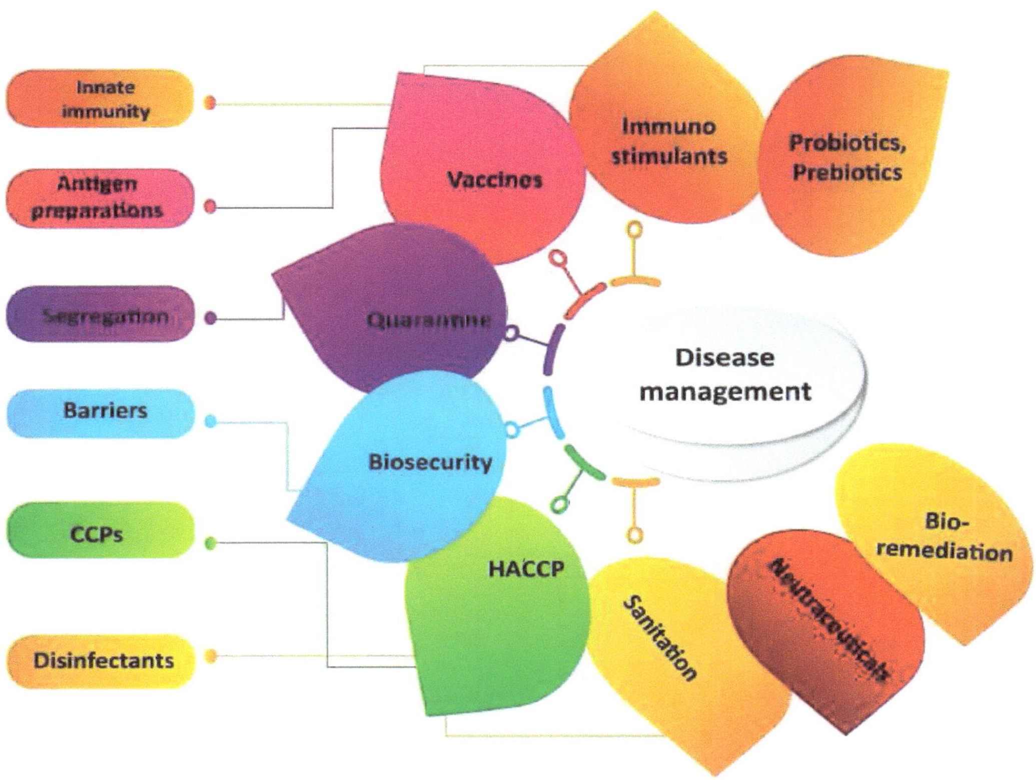

Fig. (2). Representation of various disease management strategies in a culture system.

Sanitation

Proper sanitation and disinfection strategies can be implemented at the farm to reduce the numbers of disease-causing organisms and to reduce the spread from one facility to another. The topical disinfectants used on various surfaces include formalin and benzalkonium chloride. Disinfecting methods using heat, chlorine, steam, and formalin can be used for various types of equipment [53]. Eggs can be disinfected with iodine or other solutions at the time of water hardening to reduce the incidence of disease problems in larvae. Mechanical filtration, UV light, and ozone treatments of incoming and recirculating water decrease the chance of pathogenic organisms entering the culture system. Effluent treatment of wastewater from aquaculture facilities reduces the release of microorganisms into the environment [54]. Precaution is needed with live food usage, even though fresh foods can be a high-quality supply of nutrients, these may also be a cause of pathogens. For manufactured feed, proper storage in a dry place and usage according to the manufacturer's recommendations must be ensured to prevent the loss of nutrients and the build-up of pathogens or mycotoxins [50].

HACCP Principles on the Farm

'Hazard Analysis and Critical Control Point' identifies and limits hazards on the farm. The first principle is to identify hazards, both microbiological and physicochemical, in every procedure, from stocking to harvesting [56]. The second principle is to establish 'Critical Control Points' (CCPs) at which action can be taken to reduce the hazard. There are various control points in the culture facility at which pathogen reduction can be achieved using a biosecurity programme. The third principle is to allocate 'Critical Limits' to which the hazard must be reduced. The fourth principle is 'Monitoring' which includes the observation and measurement of cleaning and disinfection processes to make sure that the critical limits are met at each step, such as source water quality, disinfection of exchanged water, equipment, and sampling kits. The fifth principle is 'Correction' in which action must be taken if the critical limits are not met at any step [56]. The sixth principle is 'Recording', which includes the collected data of the implementation of a biosecurity programme. Separate records should be maintained of products used, critical limits, and cleaning schedules. The seventh principle is 'Verification', which includes the procedures and tests to ensure the biosecurity system is working correctly. The verification process can be performed by an outside organisation or person [56].

Biosecurity

Biosecurity refers to preventing the breach, occurrence, and spread of unnecessary or harmful natural organisms or agents [57]. Biosecurity is an indispensable set of tools for the prevention, control, and eradication of infectious diseases and the prevention of human, animal, and environmental losses. Most disease occurrence results from the broad movement of live fishes and shrimps, associated with the entry and transfers of pathogens [50]. Diseases can spoil aquatic ecosystems by damaging the economy, lifestyles, and environment. Biosecurity components reduce the risk of pathogen introduction and pathogen spread to the facility; also reduce susceptibility to infection and disease; and involve careful planning of the farm facility [57]. There are three critical levels of biosecurity, such as conceptual biosecurity, structural biosecurity, and operational biosecurity. Conceptual biosecurity involves the analysis of the location of the farm, such as its proximity to other farms and to potential disease carriers [57]. Structural biosecurity concerns materials used in tanks, ponds, and building construction. Operational biosecurity concerns daily operational procedures such as effective disease monitoring for viral, bacteria, and protozoan infections; routine disinfection [50]. Advantages of keeping biosecurity include the provision of healthier fish stock, higher productivity, controlled pathogen levels, higher profits, and reduced losses in the farm [55]. There are a variety of risk assessments that can be used for aquatic animal diseases of finfishes, molluscs, and crustaceans [58].

Quarantine

Quarantine refers to maintaining a stock of aquatic animals depending upon the disease of concern, in segregation with no contact with other aquatic animals or population, to continue observation for a particular time and, if suitable, to testing and treatment, together with appropriate treatment of the wastewaters [55]. The isolation of organisms to prevent the entry or spread of infectious disease is a standard practice that is extremely important in aquaculture. Quarantine protocols include testing of a sample of shrimp or fish before bringing them onto the facility [55].

SPF, SPR, and Development of Disease Tolerant Varieties

Specific Pathogen-Free (SPF), Specific Pathogen Resistant (SPR), and Specific Pathogen Tolerant (SPT) certified stocks can be used that serve as a second line of defence for the shrimp industry for long term disease prevention [59]. SPF: The specific pathogen-free status signifies that the stocks have passed through a thorough quarantine [55], screening, and repeated examination that resolves them to be free from specified pathogens of concern. Example: *P. vannamei and P. stylirostris* SPF stocks. SPF stock is the result of an active disease surveillance program which can be procured from the certified biosecure facilities. The specifically listed pathogens must be excluded from the stock physically. After procurement of the stocks from the SPF facility, the stock may be designated as High Health (HH) animals [59]. SPR: Selective breeding of fishes can be done for higher resistance to diseases. Resistance to a specific disease is dependent on relatively simple mutations involving a few genes. The mutation may result in a change in the protein synthesis and so causes incompatibility between pathogen and host. This natural resistance may be commercially exploited without alternative genetic engineering. SPT: The vast majority of SPR stocks are being sold as SPT (Specific Pathogen Tolerance). They are tolerant of the particular pathogen up to a certain level of exposure [59].

Vaccine

To reduce the outbreak of fish diseases, vaccination of healthy cultured animals is now a common practice in temperate regions. The vaccine targets the specific immune response of the organism, hence it is suitable for fishes and not the crustaceans [14]. A univalent vaccine is planned to immunise against a single antigen, and a polyvalent vaccine is planned to immunise against two or more antigens of the pathogen or pathogens [14]. Most commercial vaccines are monovalent and confer protection against only a specific pathogen due to the complex antigenic structure of different pathogens. In some instances, a univalent vaccine may be preferred for rapidly inducing a potent immune response. Inactivated vaccines are commercially available to be used against vibriosis in the aquaculture industry. However, vaccines are not antibiotics and will not be useful for curing a disease once it has occurred [14]. Types: There are many types of vaccines available such as DNA vaccine, mRNA vaccine, recombinant protein vaccine, peptide vaccine, subunit vaccine, inactivated vaccine, and live vaccine. Killed vaccines are produced by inactivating the disease-causing microorganism using some methods including chemicals or heat [61]. Vaccines can be administered in three primary ways, such as the oral method, immersion, and injection. For developing a vaccine, the causative organism has to be identified, and scientific investigation of challenge test, disease pathology, vaccine formulation choice, commercial consideration, sterility, and safety test has to be done. Vaccines for protection against parasitic and fungal diseases are not popularised, even though laboratory experiments are ongoing [60]. RNAi technology applications for cultured shrimp: RNA interference (RNAi) serves as a tool for identifying the dsRNA (double-stranded RNA) genome components of pathogens by the innate immune defence system of crustaceans. RNAi selectively inhibits the expression of viral gene products by cleaving the nucleotide sequences [62].

Immunostimulants

In modern years, the use of immunostimulants for shrimp or finfish disease management is highly renowned for reducing the immune suppressive effects of stress [51]. Common immunostimulants include lipopolysaccharides, beta 1-3glucan [63], and peptidoglycans [51]. Categories of compounds: Quite a lot of immunostimulants have been evaluated in finfishes. Immunostimulants are classified based on their source and manner of activities, such as bacteria and bacterial products, chemical agents, plant extracts, complex carbohydrates, immunity enhancing drugs, animal extracts, nutritional factors, cytokines, and lectins. Another categorisation is specific and non-specific immunostimulants [64]. Specific immunostimulation is connected to the enhancement of the host's defence system towards a unique antigen, whereas non-specific immunostimulation is an effort to improve immunologic capabilities against several pathogens [51]. Freund's Complete Adjuvant (FCA) is a combination of *Mycobacterium butyricum* cell wall and mineral oil, which raises the immune reaction in fish. Non-specific defence mechanisms in fish and shellfish, like phagocytic activity and protection against bacterial pathogens, can be increased by yeast-derived glucans [65].

Neutraceuticals

The term 'nutraceutical was coined by Stephen DeFelice from the words 'pharmaceutical' and 'nutrition' in the year 1989. A nutraceutical is any food substance that may range from dietary supplements, isolated nutrients, genetically engineered foods, and herbal products that provide medical and health benefits, such as prevention and treatment of disease [66]. Dietary incorporation of exogenous protease such as pancreatin (porcine pancreatic extract) was found to increase the feed assimilation and growth rate significantly. Exogenous phytase, lipase, chitinase, cellulose, and carbohydrases addition is also known to improve the growth performance, nutrient utilisation, increased microbial protein synthesis, removal of fragment blockage in the intestine, hydrolysis of carboxyl ester bonds, and utilisation of unavailable dietary carbohydrate. Examples of immune-boosting nutraceuticals are microbial levan, bovine lactoferrin, nucleotide, fucoidan, propolis extract, and polysaccharides [51].

Probiotics

The probiotics, when administered, affect the host animal by improving its intestinal microbial balance by colonising the gut to minimise the pathogenic microbes through a competitive exclusion process [52]. Gastrointestinal microbiome diversity is dependent on the environment hence probiotics can be administered to the aquatic environment which is called 'water probiotics', while those administered by diet are called 'feed probiotics'. Water probiotics are also called environmental probiotics, when added directly into the water can benefit the culture system by indirectly maintaining water quality. Some examples of probiotics include *Lactobacillus*, *Vibrio sp*, *Pseudomonas sp*, *Bacillus sp.* and yeasts. They can be given to fish in the form of capsules, granules, oral paste, dispensable water powders, liquids, and directly as feed additives [67]. Bioencapsulation of probiotics can be done by enriching rotifers with probiotics. Mode of action of probiotics: Probiotics function by the production of inhibitory compounds, competition for available energy, competition for chemicals, competition for adhesion sites, inhibition of virulence genes, providing a source of nutrients, enzymatic contribution to digestion, enhancement of immune response, improvement of water quality, competition for binding sites, stimulation of epithelial barrier function, and stimulation of immune response *via* an increase of IgA and anti-inflammatory cytokines [67].

Prebiotics

Prebiotics are non-digestible food ingredients used for enhancing the health-promoting bacteria, which include dietary carbohydrates, peptides, proteins, and specific lipids [68]. The carbohydrates derived from plant and yeast origins, such as inulin, fructose oligosaccharides, Transgalacto Oligosaccharides (TOS), lactulose, Isomalto Oligosaccharides (IMO), Mannan Oligosaccharide (MOS), lactosucrose, Xylooligosaccharides (XOS), soybean oligosaccharides and glucooligosaccharides can be used as potential prebiotics. They function to increase the uptake of glucose and the bioavailability of trace elements. The beneficial health effects of prebiotics include promoting the growth of probiotic bacteria and suppressing the effects of harmful bacteria [68].

Bioremediation

Bioremediation is the process of degrading the environmental contaminants into less toxic forms using living organisms. It is the process of using naturally occurring bacteria and fungi or plants to degrade or detoxify substances hazardous to the environment [69]. It is effective only where environmental conditions permit microbial growth and activity to attack the pollutants, converting them into harmless products enzymatically [70]. The process can improve water quality by nitrification and the removal of toxic substances, such as hydrogen sulfide, ammonia, and nitrite from the water [69]. Microbial diversity used for the remediation process include *Aeromonas, Pseudomonas, Mycobacteria, Bacilli, Arthrobacter, Streptomyces, and Cyanobacteria*. Polychlorinated biphenyls and chloroform can be remediated using anaerobic bacteria. A diverse range of persistent environmental pollutants can be degraded using ligninolytic fungi [70].

Advantages of Bioremediation: It is a natural, eco-friendly process and destroys pollutants rather than transforming them into other compounds. It works on a variety of pollutants, such as organic and inorganic compounds. It is easy to implement and low cost compared to other treatments [70].

DISEASE DIAGNOSIS METHODS

Successful diagnosis of a disease-causing agent is a prerequisite for providing the correct treatment for a disease. Specific behavioral changes of the fish lik loss of feed intake, laboured respiration, surface-gasping, loss of stability, and jumping can be used for on-farm disease prediction approaches [71]. Hence, the pathoanatomic examination of the fish should be focused on fin erosion, cuts, and lesions, reddening of the skin, raised scales, presence of parasites, damaged gills, opercular distension, trailing mucus, and the presence of lumps. The water quality examination should also be done to assess the correctness of the environmental parameters [71].

Surveillance and Monitoring

Surveillance and monitoring are the constant examination of a specified fish culture facility or culture area to identify the existence of pathogens for control measures, which may encompass testing of a fragment of the population for the detection of changes in the prevalence of disease in a given population and its environment [72].

Controlling the Spread of Infection

The spread of infectious diseases is controlled by disinfectants such as hydrogen peroxide and malachite green; antibiotics such as sulfonamides and tetracyclines; and anthelmintic drugs such as pyrethroids and avermectins. The effectiveness of a given chemical depends on various factors [54]. The ideal properties of disinfectants and sanitisers include high penetrability, a broad spectrum of antibacterial activity, lack of irritation, low toxicity, non-interference with normal healing processes, cost- effectiveness, non-corrosiveness, non-staining ability, activity in necrotic tissues, and a higher degree of stability [53]. The ability to penetrate the crevices, cavities, and films of organic matter and maintain a lethal concentration of the agent is essential to maintain the bactericidal effect in soil and faecal material. The disinfectants may be classified as oxidising agents such as halogens, peroxides, and potassium permanganate; reducing agents such as formaldehyde; and detergents such as quaternary ammonium compounds [50].

Different Types of Treatment

The drugs can be administered by waterborne, oral, and injection routes. The waterborne treatment methods are categorised into the dip, bath, flush, constant flow, and prolonged immersion. Dip treatment is of short duration such as 15-60 seconds with the use of a high concentration of the chemical [54]. In bath treatment, the fishes are exposed to a high concentration of the drug for specified short periods ranging from a few minutes to several hours [73]. The flush treatment is a modification of bath treatment in which a high concentration of the drug is used with constant water flow. The entire dose can be flushed in a few minutes [73]. By using the constant flow treatment method, chemicals such as formalin and malachite green are commonly delivered. A slow flow rate is preferred, and the drug is allowed to fall into the upstream at a rate depending on the required concentration. For oral medication, the drugs are mixed with the pelleted feeds at the required dosage. This method can be used mainly for caged and pond reared fishes. Oral medication can be carried out by using commercial medicated feeds, artemia loaded with medicine [74], or by preparation of medicated feed.

Antibiotics

Antimicrobial agents are often defined as substances that can kill or inhibit the growth of microorganisms. Antibiotics are produced by microorganisms or derived synthetically and selectively inhibit the growth of others and have no effect on viral infections. Use is often categorised as therapeutic, prophylactic or metaphylactic [75]. The use of antimicrobials to prevent disease in either individuals or groups is called prophylaxis. In the tank or cage fish culture facilities, the therapeutic antibiotics are administered for brief periods *via* the oral route through incorporated feeds [3]. Examples of antibiotics include florfenicol, oxolinic acid, chloramphenicol, OTC, nitrofurans, potentiated sulfonamide, quinolones, and flumequine. In general, antibiotics work by one of two mechanisms: bactericidal and bacteriostatic. A bacteriostatic effect results in the reduction of bacterial multiplication by interference with DNA replication, protein production, or cellular metabolism. Examples include tetracyclines, sulfonamides, chloramphenicol, and macrolides [75]. Specific vitamin synthesis pathways are inhibited by antibiotics such as trimethoprim and sulfonamide, which ultimately inhibits DNA synthesis. Polymyxins accumulate within the bacterial cell wall and exert their inhibitory effects by increasing bacterial membrane permeability. The cyclic lipopeptide daptomycin inserts its lipid tail into the bacterial cell wall, causing membrane depolarisation and, eventually, the death of the bacterium [75]. Uncontrolled and unregulated use of antibiotics results in the development and spread of antibiotic resistance, which reduces the efficacy of antibiotic treatments, causes residue accumulations in tissues and potential environmental hazards, and raises public health issues [76]. Multidrug resistance is found in bacteria. Multiresistant *Aeromonas salmonicida* isolates are described in various parts of the world, and transferable resistance plasmids are commonly detected in these strains [77]. Only a few antibiotics are licensed to be used in aquaculture with strict regulations. The antibiotic-resistant bacteria occurrence in foods of animal origin is a vital health threat because these pathogens may not respond to the specific antibiotic treatments [78]. The use of pharmacological agents, antibiotics, and other chemical agents should be considered a method of the last resort in shrimp farming and aquaculture generally. Hence, alternative treatments using essential oils and phage therapy can be used [78].

The primary role of essential oils is believed to be a defence against pathogens due to their antibacterial, antiviral, antifungal, and insecticidal effects; they also contain phenolic compounds of antimicrobial nature [79]. Bacteriophages have some specific advantages compared to therapeutic agents, such as they need specific narrow host ranges, meaning that they do not harm the traditional intestinal microbiota; they can make multiple copies in the bacterial target, which reduces the need for multiple administrations; and there have been no reports of side effects [80].

Treatments of Parasitic Infections

Various chemicals can be used for the treatment of parasites, such as the use of ivermectin for isopods and copepods; formalin against internal protozoan infections [81]; levamisole against monogenean infections; and fumagillin against external protozoan infections. Paratransgenic Artemia can be developed as a platform for the control of infectious parasitic diseases in shrimp mariculture [74]. Expression of molecules of infectious agents in feed organisms such as Artemia and algae can benefit mariculture by offering a powerful tool against infectious organisms [81].

CONCLUDING REMARKS

Healthiness supervision in mariculture advances production and diminishes the threat associated with disease. Health management is a collective responsibility in mariculture, and each participant's contribution is essential for the management process. For sustainable mariculture, the components of concern in disease management are disease diagnostics, monitoring, and management. The diagnosis must be rapid, specific, and sensitive, and monitoring is an integral component. The management measures improve productivity and reduce the risk associated with the diseases. Science-based health management practices, concerned with seasonal factors and crop planning, can be carried out. Health management can be done as a collective measure at the national level and within each mariculture facility at a system level. The national-level management strategies include managing trans-boundary aquatic animal movement, quarantine, providing an aquatic animal health information system, contingency planning, surveillance techniques, import risk analysis, research, and development. The system-level best management strategies include SPF, SPR, and SPT usage, periodic environmental and health monitoring, and the use of approved inputs (feed, probiotics, drugs, immunostimulants, vaccines, and chemicals).

REFERENCES

[1] Suto SH. Mariculture of seaweeds and its problems in Japan. NOAA Tech Rep NMFS Circ 1974; 388: 7-16.

[2] Klochkova TA, Shim JB, Hwang MS, Kim GH. Host–parasite interactions and host species susceptibility of the marine oomycete parasite, Olpidiopsis sp., from Korea that infects red algae. J Appl Phycol 2012; 24(1): 135-44. [http://dx.doi.org/10.1007/s10811-011-9661-8]

[3] Woo PT, Gregory DW, Eds. Diseases and disorders of finfish in cage culture. CABI 2014. [http://dx.doi.org/10.1079/9781780642079.0000]

[4] Skov J, Mehrdana F, Marana MH, *et al.* Parasite infections of rainbow trout (Oncorhynchus mykiss) from Danish mariculture. Aquaculture 2014; 434: 486-92. [http://dx.doi.org/10.1016/j.aquaculture.2014.08.041]

[5] Timi JT, MacKenzie K. Parasites in fisheries and mariculture. Parasitology 2015; 142(1): 1-4. [http://dx.doi.org/10.1017/S0031182014001188] [PMID: 25630219]

[6] Sindermann CJ. 1971.

[7] Rottmann RW, Francis-Floyd R, Durborow R. The role of stress in fish disease. Stoneville, MS: Southern Regional Aquaculture Center 1992.

[8] Kusuda R, Salati F. Major bacterial diseases affecting mariculture in Japan. Annu Rev Fish Dis 1993; 3: 69-85. [http://dx.doi.org/10.1016/0959-8030(93)90029-B]

[9] Toranzo AE, Devesa S, Heinen P, Riaza A, Nunez S, Barja JL. Streptococcosis in cultured turbot caused by an Enterococcus-like bacterium. Bull Eur Assoc Fish Pathol 1994; 14: 19-9.

[10] Toranzo AE, Magariños B, Romalde JL. A review of the main bacterial fish diseases in mariculture systems. Aquaculture 2005; 246(1-4): 37-61. [http://dx.doi.org/10.1016/j.aquaculture.2005.01.002]

[11] Zeng Y, Ma Y, Wei C, *et al.* Bacterial diversity in various coastal mariculture ponds in Southeast China and in diseased eels as revealed by culture and culture-independent molecular techniques. Aquacult Res 2010; 41(9): e172-86. [http://dx.doi.org/10.1111/j.1365-2109.2010.02499.x]

[12] Li G, Zhao D, Huang L, *et al.* Identification and phylogenetic analysis of Vibrio vulnificus isolated from diseased Trachinotus ovatus in cage mariculture. Aquaculture 2006; 261(1): 17-25. [http://dx.doi.org/10.1016/j.aquaculture.2006.07.013]

[13] Hennersdorf P, Kleinertz S, Theisen S, *et al.* Microbial diversity and parasitic load in tropical fish of different environmental conditions. PLoS One 2016; 11(3)e0151594 [http://dx.doi.org/10.1371/journal.pone.0151594] [PMID: 27018789]

[14] Ruixuan WA, Liwen XU, Juan FE. A review of the pathogenic bacteria, diagnosis and vaccine of principal bacterial diseases of mariculture fish. Southern Fisheries Science 2005; 1(6): 72-9.

[15] Rigos G, Katharios P. Pathological obstacles of newly-introduced fish species in Mediterranean mariculture: a review. Rev Fish Biol Fish 2010; 20(1): 47-70. [http://dx.doi.org/10.1007/s11160-009-9120-7]

[16] Zhang XH, He X, Austin B. *Vibrio harveyi*: a serious pathogen of fish and invertebrates in mariculture. Mar Life Sci Technol 2020; 2(3): 231-45. [http://dx.doi.org/10.1007/s42995-020-00037-z] [PMID: 32419972]

[17] Cai SH, Lu YS, Wu ZH, Jian JC, Wang B, Huang YC. Loop-mediated isothermal amplification method for rapid detection of *Vibrio alginolyticus*, the causative agent of vibriosis in mariculture fish. Lett Appl Microbiol 2010; 50(5): 480-5. [http://dx.doi.org/10.1111/j.1472-765X.2010.02823.x] [PMID: 20214730]

[18] Saravanan K, Nilavan SE, Sudhagar SA, Naveenchandru V. Diseases of mariculture finfish species: a review 2003.

[19] Ina-Salwany MY, Al-saari N, Mohamad A, *et al.* Vibriosis in fish: a review on disease development and prevention. J Aquat Anim Health 2019; 31(1): 3-22. [http://dx.doi.org/10.1002/aah.10045] [PMID: 30246889]

[20] Zhao DH, Sun JJ, Liu L, *et al.* Characterization of two phenotypes of Photobacterium damselae subsp. damselae isolated from diseased juvenile Trachinotus ovatus reared in cage mariculture. J World Aquacult Soc 2009; 40(2): 281-9. [http://dx.doi.org/10.1111/j.1749-7345.2009.00251.x]

[21] Pedersen K, Skall HF, Lassen-Nielsen AM, Bjerrum L, Olesen NJ. *Photobacterium damselae* subsp. *damselae*, an emerging pathogen in Danish rainbow trout, *Oncorhynchus mykiss* (Walbaum), mariculture. J Fish Dis 2009; 32(5): 465-72. [http://dx.doi.org/10.1111/j.1365-2761.2009.01041.x] [PMID: 19364386]

[22] Austin, B. Bacterial pathogens of marine fish, In Oceans and health: pathogens in the marine environment. Springer 2005, Boston, MA, pp. 391-413

[23] Xiao J, Wang Q, Liu Q, Wang X, Liu H, Zhang Y. Isolation and identification of fish pathogen *Edwardsiella tarda* from mariculture in China. Aquacult Res 2008; 40(1): 13-7. [http://dx.doi.org/10.1111/j.1365-2109.2008.02101.x]

[24] Egusa S. Disease problems in Japanese yellowtail, Seriola quinqueradiata, culture: a review Diseases of commercially important marine fish and shellfish 1983; 182: 10-8.

[25] Skall HF, Olesen NJ, Mellergaard S. Viral haemorrhagic septicaemia virus in marine fish and its implications for fish farming – a review. J Fish Dis 2005; 28(9): 509-29. [http://dx.doi.org/10.1111/j.1365-2761.2005.00654.x] [PMID: 16266325]

[26] Walker PJ, Winton JR. Emerging viral diseases of fish and shrimp. Vet Res 2010; 41(6): 51. [http://dx.doi.org/10.1051/vetres/2010022] [PMID: 20409453]

[27] Zorriehzahra MJ, Hassantabar F, Ziarati M, *et al.* Impact of Viral Nervous Necrosis (VNN) Disease as a New Threat to Global Fisheries and Aquaculture Development, a Review. Iran J Virol 2019; 13(2): 42-57.

[28] Bandín I, Souto S. Betanodavirus and VER disease: a 30-year research review. Pathogens 2020; 9(2): 106. [http://dx.doi.org/10.3390/pathogens9020106] [PMID: 32050492]

[29] Arulmoorthy MP, Anandajothi E, Vasudevan S, Suresh E. Major viral diseases in culturable penaeid shrimps: a review. Aquacult Int 2020; 28(5): 1939-67. [http://dx.doi.org/10.1007/s10499-020-00568-3]

[30] Sweet MJ, Bateman KS. Reprint of 'Diseases in marine invertebrates associated with mariculture and commercial fisheries'. J Sea Res 2016; 113: 28-44. [http://dx.doi.org/10.1016/j.seares.2016.06.001] [PMID: 32336937]

[31] Liao X, He J, Li C. Decapod iridescent virus 1: An emerging viral pathogen in aquaculture. Rev Aquac 2022.

[32] Nagasawa K. Parasitic copepods of marine fish cultured in Japan: a review. J Nat Hist 2015; 49(45-48): 2891-903. [http://dx.doi.org/10.1080/00222933.2015.1022615]

[33] Paperna I, Baudin Laurencin F. Parasitic infections of sea bass, Dicentrarchus labrax, and gilt head sea bream, Sparus aurata, in mariculture facilities in France. Aquaculture 1979; 16(2): 173-5. [http://dx.doi.org/10.1016/0044-8486(79)90148-0]

[34] Palm HW, Kleinertz S, Rückert S. Parasite diversity as an indicator of environmental change? An example from tropical grouper (*Epinephelus fuscoguttatus*) mariculture in Indonesia. Parasitology

2011; 138(13): 1793-803.
[http://dx.doi.org/10.1017/S0031182011000011] [PMID: 21320385]

[35] Zavyalova EA, Alontseva DA, Belimenko VV, Bulina KY, Droshnev AE. Lepeophtiriosis caused by Lepeophtheirus salmonis in a mariculture of Russia and the world. InIOP Conference Series: Earth and Environmental Science; (Vol. 421, No. 8, p. 082005). IOP Publishing 2020.
[http://dx.doi.org/10.1088/1755-1315/421/8/082005]

[36] Power C, Nowak BF, Cribb TH, Bott NJ. Bloody flukes: a review of aporocotylids as parasites of cultured marine fishes. Int J Parasitol 2020; 50(10-11): 743-53.
[http://dx.doi.org/10.1016/j.ijpara.2020.04.008] [PMID: 32619429]

[37] Afonso-Dias, I, Menezes, GM, Mackenzie, K and Eiras, JC. Applied Aspects of Marine Parasitology. Proceedings of the International Workshop on Marine Parasitology; Universidade dos Açores. 2008.

[38] Nowak BF. Parasitic diseases in marine cage culture – An example of experimental evolution of parasites? Int J Parasitol 2007; 37(6): 581-8.
[http://dx.doi.org/10.1016/j.ijpara.2007.01.003] [PMID: 17316650]

[39] MacKenzie K, Kalavati C, Gaard M, Hemmingsen W. Myxosporean gall bladder parasites of gadid fishes in the North Atlantic: Their geographical distributions and an assessment of their economic importance in fisheries and mariculture. Fish Res 2005; 76(3): 454-65.
[http://dx.doi.org/10.1016/j.fishres.2005.07.014]

[40] Rückert S, Palm HW, Klimpel S. Parasite fauna of seabass (Lates calcarifer) under mariculture conditions in Lampung Bay, Indonesia. J Appl Ichthyology 2008; 24(3): 321-7.
[http://dx.doi.org/10.1111/j.1439-0426.2008.01064.x]

[41] Brooker AJ, Shinn AP, Bron JE. A review of the biology of the parasitic copepod Lernaeocera branchialis (L., 1767) (Copepoda: Pennellidae). Adv Parasitol 2007; 65: 297-341.
[http://dx.doi.org/10.1016/S0065-308X(07)65005-2] [PMID: 18063099]

[42] Yuniar AT, Palm HW, Walter T. Crustacean fish parasites from Segara Anakan Lagoon, Java, Indonesia. Parasitol Res 2007; 100(6): 1193-204.
[http://dx.doi.org/10.1007/s00436-006-0391-9] [PMID: 17219222]

[43] Ramaiah N. A review on fungal diseases of algae, marine fishes, shrimps and corals 2006.

[44] Comrnittee, M. The fungal pathogen ichthyophonus hoferi in sea herring, Clupea harengus: a perspective from the western north atlantic.

[45] Delroisse J, Van Wayneberghe K, Flammang P, *et al.* Epidemiology of a SKin Ulceration Disease (SKUD) in the sea cucumber Holothuria scabra with a review on the SKUDs in Holothuroidea (Echinodermata). Sci Rep 2020; 10(1): 22150.
[http://dx.doi.org/10.1038/s41598-020-78876-0] [PMID: 33335179]

[46] Shields JD. Diseases of spiny lobsters: A review. J Invertebr Pathol 2011; 106(1): 79-91.
[http://dx.doi.org/10.1016/j.jip.2010.09.015] [PMID: 21215357]

[47] Fire, SE and Van Dolah, FM (2012), Marine biotoxins: emergence of harmful algal blooms as health threats to marine wildlife.

[48] Couch JA. Prospective study of infectious and noninfectious diseases in oysters and fishes in three Gulf of Mexico estuaries. Dis Aquat Organ 1985; 1: 59-82.
[http://dx.doi.org/10.3354/dao001059]

[49] Hadfield CA. Noninfectious Diseases (Environmental). Clinical Guide to Fish Medicine 2021; 13: 357-77.
[http://dx.doi.org/10.1002/9781119259886.ch22]

[50] McVicar AH. Epidemiology/epizootiology: a basis for control of disease in mariculture. Pathology in marine science 1990; 397-405.

[51] Ringø E, Olsen RE, Vecino JG, Wadsworth S, Song SK. Use of immunostimulants and nucleotides in

aquaculture: a review. J Mar Sci Res Dev 2012; 2(1): 104.

[52] Hoseinifar SH, Sun YZ, Wang A, Zhou Z. Probiotics as means of diseases control in aquaculture, a review of current knowledge and future perspectives. Front Microbiol 2018; 9: 2429. [http://dx.doi.org/10.3389/fmicb.2018.02429] [PMID: 30369918]

[53] Park SI. Disease control in Korean aquaculture. Fish Pathol 2009; 44(1): 19-23. [http://dx.doi.org/10.3147/jsfp.44.19]

[54] Brock JA, Bullis R. Disease prevention and control for gametes and embryos of fish and marine shrimp. Aquaculture 2001; 197(1-4): 137-59. [http://dx.doi.org/10.1016/S0044-8486(01)00585-3]

[55] Subasinghe R. Fish health and quarantine. FAO Fish Circ 1997; 45-9.

[56] Reilly A, Käferstein F. Food safety hazards and the application of the principles of the hazard analysis and critical control point (HACCP) system for their control in aquaculture production. Aquacult Res 1997; 28(10): 735-52. [http://dx.doi.org/10.1111/j.1365-2109.1997.tb00999.x]

[57] Weirich CR, Segars AL, Bruce JE, Browdy CL. Development and implementation of biosecurity protocols and procedures at the Waddell Mariculture Center Biosecurity in aquaculture production systems: exclusion of pathogens and other undesirables Aquaculture Interchange Program. Waimanalo, Hawaii: Oceanic Institute 2003; pp. 139-56.

[58] Oidtmann BC, Thrush MA, Denham KL, Peeler EJ. International and national biosecurity strategies in aquatic animal health. Aquaculture 2011; 320(1-2): 22-33. [http://dx.doi.org/10.1016/j.aquaculture.2011.07.032]

[59] Sanz VA. Specific pathogen free (SPF), specific pathogen resistant (SPR) and specific pathogen tolerant (SPT) as part of the biosecurity strategy for whiteleg shrimp (Penaeus vannamei Boone 1931). Asian Fish Sci 2018; 31: 112-20.

[60] Adams A. Progress, challenges and opportunities in fish vaccine development. Fish Shellfish Immunol 2019; 90: 210-4. [http://dx.doi.org/10.1016/j.fsi.2019.04.066] [PMID: 31039441]

[61] Muktar Y, Tesfaye S, Tesfaye B. Present status and future prospects of fish vaccination: a review. J Vet Sci Technol 2016; 7(2): 299. [http://dx.doi.org/10.4172/2157-7579.1000299]

[62] Nguyen DV, Christiaens O, Bossier P, Smagghe G. RNA interference in shrimp and potential applications in aquaculture. Rev Aquacult 2018; 10(3): 573-84. [http://dx.doi.org/10.1111/raq.12187]

[63] Mohan K, Ravichandran S, Muralisankar T, *et al.* Potential uses of fungal polysaccharides as immunostimulants in fish and shrimp aquaculture: A review. Aquaculture 2019; 500: 250-63. [http://dx.doi.org/10.1016/j.aquaculture.2018.10.023]

[64] Mastan SA. Use of immunostimulants in aquaculture disease management. Int j fish aquat sci 2015; 2(4): 277-80.

[65] Zhenming C, Zhiqiang L, Lingmei G, *et al.* Marine yeasts and their applications in mariculture. J Ocean Univ China 2006; 5(3): 251-6. [http://dx.doi.org/10.1007/s11802-006-0010-5]

[66] Varghese T. SanalEbeneeza SA, Pal AK. Mitigation of stress in fish through nutraceuticals. Development 2021; 3(1): 1014.

[67] Sharifuzzaman SM, Austin B. Probiotics for disease control in aquaculture. Diagnosis and control of diseases of fish and shellfish. 2017; 7: 189-222. [http://dx.doi.org/10.1002/9781119152125.ch8]

[68] Loh JY, Chan HK, Yam HC, In LLA, Lim CSY. An overview of the immunomodulatory effects

exerted by probiotics and prebiotics in grouper fish. Aquacult Int 2020; 28(2): 729-50. [http://dx.doi.org/10.1007/s10499-019-00491-2]

[69] Antony SP, Philip R. Bioremediation in shrimp culture systems. Naga The WorldFish Center Quarterly 2006; 29(3 & 4): 62-6.

[70] Hu WJ, Yang SY, Zhu XM. The impact of mariculture on the marine ecosystem and studies on bioremediation. Nat Sci Soc 2007; 1: 197-202.

[71] Cunningham CO. Molecular diagnosis of fish and shellfish diseases: present status and potential use in disease control. Aquaculture 2002; 206(1-2): 19-55. [http://dx.doi.org/10.1016/S0044-8486(01)00864-X]

[72] Sahu BB, Radheyshy AM, Kumar K, Mukherjee SC, Ayyappan S. Farmer participatory fish disease surveillance and monitoring: use of PRA tools. Trop Agric Res Ext 1999; 2: 118-23.

[73] Sudova E, Machova J, Svobodova Z, Vesely T. Negative effects of malachite green and possibilities of its replacement in the treatment of fish eggs and fish: a review. Vet Med (Praha) 2007; 52(12): 527-39. [http://dx.doi.org/10.17221/2027-VETMED]

[74] Subhadra B, Hurwitz I, Fieck A, Rao DVS, Subba Rao G, Durvasula R. Development of paratransgenic *Artemia* as a platform for control of infectious diseases in shrimp mariculture. J Appl Microbiol 2010; 108(3): 831-40. [http://dx.doi.org/10.1111/j.1365-2672.2009.04479.x] [PMID: 19702854]

[75] Vignesh R, Karthikeyan BS, Periyasamy N, Devanathan K. Antibiotics in aquaculture: an overview. South Asian J Exp Biol 2011; 1(3): 114-20. [http://dx.doi.org/10.38150/sajeb.1(3).p114-120]

[76] Musefiu TA, Olasunkanmi SM. Antibiotics resistance in bacteria strains isolated from fish: potential health risk. J Biol Agric Healthc 2015; 10(5)

[77] Deng Y, Xu L, Chen H, *et al.* Prevalence, virulence genes, and antimicrobial resistance of Vibrio species isolated from diseased marine fish in South China. Sci Rep 2020; 10(1): 14329. [http://dx.doi.org/10.1038/s41598-020-71288-0] [PMID: 32868874]

[78] Singha J, Abraham TJ, Roy A, T Dash G, Nagesh TS, Patil PK. Use of aquadrugs, chemicals and biological products in freshwater aquaculture systems of West Bengal, India.

[79] Dawood MAO, El Basuini MF, Zaineldin AI, *et al.* Antiparasitic and antibacterial functionality of essential oils: An alternative approach for sustainable aquaculture. Pathogens 2021; 10(2): 185. [http://dx.doi.org/10.3390/pathogens10020185] [PMID: 33572193]

[80] Ninawe AS, Sivasankari S, Ramasamy P, Kiran GS, Selvin J. Bacteriophages for aquaculture disease control. Aquacult Int 2020; 28(5): 1925-38. [http://dx.doi.org/10.1007/s10499-020-00567-4]

[81] Athanassopoulou F, Pappas IS, Bitchava K. An overview of the treatments for parasitic disease in Mediterranean aquaculture. Options Méditerr Sér A: Sémin Méditerr 2009; 86: 65-83.

CHAPTER 6

Developmental Trends in Aquafeed for Mariculture Species

Reshma Devi Ramesh[1,*], Revathi Nandhakumar[1], Gowri Ayyadurai[1], Sudharsan Parthasarathy[1], Koperuncholan Marimuthu[2], K Gobalan[3] and **Siva Vijayakumar Tharmasivam[1]**

[1] *Department of Biotechnology, Srimad Andavan Arts and Science College, Tiruchirappalli, Tamil Nadu, India*

[2] *Department of Botany, Srimad Andavan arts and Science College, Tiruchirappalli, Tamil Nadu, India*

[3] *Department of Biotechnology, Jamal Mohamed College, Tiruchirappalli, Tamil Nadu, India*

Abstract: Fish play a crucial role in ensuring global food security, particularly in developing countries, where the demand for additional protein sources continues to rise alongside population growth. Aquaculture, the fastest-growing sector in food production, focuses on the breeding and rearing of fish and other aquatic species. Within this industry, fish nutrition and feeding are critical for sustainable growth and overall productivity. The use of feed ingredients that support growth while providing a balanced supply of essential nutrients is fundamental to maintaining healthy fish and, by extension, healthy consumers. Scientific research has significantly advanced the development and efficacy of aquafeeds, ensuring that both farmed and wild fish receive optimal levels of essential nutrients, including vitamins, fatty acids, and amino acids, to maximize growth, health, and survival. Freshly formed resources include macroalgae, transgenic-raised, HUFA-producing plants, and microorganism biomass. Newly explored feed resources include herbivorous zooplankton, such as Antarctic krill, and various plant-based ingredients. Incorporating these resources can help shift carnivorous species toward lower trophic-level diets, thereby enhancing production efficiency and product quality. While these ingredients contribute, to some extent, to human nutrition, the long-term goal of mariculture in the 21st century is to rear carnivorous fish primarily on plant-based resources. The aquaculture industry has significant potential to reduce its reliance on conventional fishmeal. Improvements in feed composition and nutritional efficiency can lead to higher fish yields, lower feed costs, and reduced waste, ultimately decreasing the environmental nutrient load associated with intensive fish farming.

Keywords: Aquaculture, Aquafeed, Fish farming, HUFA, Mariculture.

* **Corresponding author Reshma Devi Ramesh:** Department of Biotechnology, Srimad Andavan Arts and Science College, Tiruchirappalli, Tamil Nadu, India; E-mail: reshmadevimb@gmail.com

G. Chelladurai, M. Paripooranaselvi & B. Makesh Kumar (Eds.)

INTRODUCTION

Aquaculture is expanding rapidly worldwide, driving increased interest in the use of compound aquafeeds and alternative external feed inputs. Changes in feed ingredients, marketing strategies, and production technologies represent key structural developments necessary for the sustainable growth of the aquaculture industry [1]. The conversion of agriculturally derived raw materials into feed is primarily managed by the aquafeed sector, which plays a critical role both nutritionally and economically. These feed units serve as the main source of animal and plant proteins essential for the proper growth of cultured species. A wide range of professionals is involved in this sector, including process engineers, economists, marketing specialists, aquaculture scientists, regulatory experts, quality control technicians, and logistics personnel. Since feed accounts for 50-60% of the total production cost in aquaculture, even small reductions in feed prices can significantly impact overall costs and profitability. Most feed ingredients have inherent nutritional and functional limitations and cannot be used in isolation at high inclusion levels for most cultured species. Therefore, aquafeed formulation requires the careful combination of multiple ingredients. To maintain profitability, farmers often rely on locally available raw material substitutes when preparing aquafeeds. However, factors such as ingredient availability, seasonal production, market prices, and processing capabilities present challenges to feed formulation. As a result, designing an optimal aquafeed formula is a critical step in the successful development of the aquaculture sector.

Since 1950, the human population and global food demand have increased dramatically. Agricultural production expanded from over 3,000 million metric tonnes in 1961 to roughly 7,000 million metric tonnes in 2010, growing at a rate slightly faster than population growth. However, land use for agriculture increased by only about 10%, from approximately 4,500 million hectares in 1961 to around 5,000 million hectares in 2010, with the primary driver of production growth being agricultural intensification [2]. Fisheries play a critical role in the global food system, providing about 2% of the world's protein consumption. As the global population continues to rise, so does the demand for fish products. Overfishing has become a major concern, and production from global capture fisheries has fluctuated annually without showing significant growth since the mid-1980s.

Currently, 8-9% of the global human protein intake from animal sources is supplied by aquaculture, which also provides nearly half of the world's fisheries production for human consumption. By the end of the century, the global population is projected to exceed ten billion, up from the current estimate of approximately 7.2 billion, with most of this growth occurring over the next forty years. As capture fisheries are not expected to expand—and may even decline—aquaculture production will need to meet the increasing demand for fish and other aquatic products. In its simplest form, animal aquaculture produces shrimp, fish, or other aquatic species that rely on naturally available food sources. However, natural productivity in these systems is often limited, so industrial chemical fertilizers or organic waste by-products are commonly used to enhance the availability of food for cultured animals.

The use of high-quality, formulated feed that is readily consumed by cultured animals can substantially increase productivity. When mechanical aeration is applied to supplement existing sources of dissolved oxygen, both feed utilization and overall production are further enhanced. For example, penaeid shrimp yields in unfertilized ponds are typically 200-300 kg/ha, whereas yields in fertilized ponds can range from 400 to 800 kg/ha. Feeding alone can increase shrimp yields to 1500-2000 kg/ha, provided dissolved oxygen levels do not become limiting. In ponds equipped with both feeding and mechanical aeration, yields can reach up to 10,000 kg/ha [3, 4]. Aquaculture often faces limitations in land and water availability, which has driven a trend toward intensification, similar to practices observed in terrestrial agriculture. For instance, farm yields of ictalurid catfish in the southern United States averaged around 1,500-1,800 kg/ha in the 1960s, gradually increasing to approximately 5,000 kg/ha by 2012. This growth has been driven by improved feed formulations, higher feeding rates, and the development of more efficient aeration systems [5].

A feed input was employed in approximately one-third of aquaculture production systems in 2012, up from about one-third of systems in 1980, according to the UN (2012). The production of aquaculture feeds reached 34.4 million tonnes in 2012 [6]. Although this is small compared to ruminant and poultry feed, which totaled 418 million tonnes, global production of fish through aquaculture and capture is significant. Feed production for aquaculture (34.4 million tonnes) and for even-toed ungulates (218 million tonnes) represents an important industry with substantial potential for future growth. Aquaculture feed contains similar components to feed for other animals, with the majority of ingredients derived from capture fisheries or agriculture, which are then processed before being sent to a feed mill for aquafeed production. Additionally, some ingredients are by-products or leftovers from fishing and agricultural activities.

FEED INGREDIENTS

Utilizing locally available raw materials as ingredients in aquaculture feed enhances resource efficiency and supports the potential for increased aquaculture production with minimal environmental impact. The evaluation of feed ingredients is also critical for feed development. It is important to differentiate the effects of feed intake, nutrient utilization, growth performance, and other metabolic processes [7]. Protein (amino acid) sources, energy sources, essential lipid sources, vitamins, mineral supplements, and specialty ingredients contribute to growth, pigmentation, sexual development, or influence the physical quality, taste, and shelf-life of the feed, and are commonly included in practical aquaculture diets [8]. Cultivation feeds are often made from small amounts of fish, fish processing waste, leftover meat, cereals, plant meals, and even organic fertilizer. However, the use of these materials as feeds presents challenges related to availability, nutritional value, and sanitary standards. The nutritional requirements of specific species are considered when formulating commercial feeds. These feeds are composed of a variety of high-quality ingredients, including fishmeal, squid meal, plant meals, ground corn flour, rice flour, meat scrap meal, feather meal, and bone meal, as well as distiller's dried solubles, fish oil, edible fats, vitamins, mineral supplements, antioxidants, and other additives. Feed is typically manufactured by adjusting particle size to suit different species and growth stages for which it is intended. The quantitative proportion of fishmeal to plant meal in feed varies by species: it is highest for salmon and trout, moderate for shrimp, and lowest for carp and catfish. Considerable efforts are being made to reduce the amount of fishmeal included in aquaculture feeds, and some manufacturers now offer feeds that contain no fishmeal at all—for example, feeds for ictalurid catfish often do not contain any fishmeal [9]. Research has demonstrated that many species can be successfully cultivated on diets consisting solely of plant proteins, although fishmeal-free feeds often include a rendered animal product as a source of animal protein (Fig. **1**).

For some saprophytic or omnivorous species, the crude protein (nitrogen × 6.25) content of aquaculture diets ranges from approximately 25% to as high as 55% for certain carnivorous species and juvenile stages. On average, most feeds contain around 35% protein. Aquaculture feeds also include lipids and other ingredients to provide energy and essential fatty acids, with common sources being edible fats, fish oil, other marine oils, or a combination of these. Compared to warm-water or tropical species such as catfish, tilapia, and marine shrimp, feeds for cold-water species like salmon and trout generally have a higher lipid content, particularly from animal oil. Similarly, each component in a feed formulation has its own advantages and limitations. The benefits of feed ingredients are generally associated with their nutritional contributions and the availability of essential nutrients, such as amino acids and lipids. Conversely, the drawbacks are related to the presence of antinutritional factors, contaminants, mold, potential mycotoxin production, low or variable quality, poor palatability, susceptibility to oxidation, cost, availability, and digestibility. With the notable exception of soybean meal, plant protein sources are typically nutritionally deficient in certain essential amino acids, which are the primary limiting proteins in cereals and other plant products. For most commercially important aquaculture species, particularly carnivorous ones, plant-based diets may fail to meet essential amino acid requirements unless supplemented with animal protein sources or crystalline amino acids. In contrast, animal protein components are commonly used to enhance the protein composition of finished feeds and to balance the amino acid profile of diets. For fish and shrimp feeds, manufacturers in various countries ensure that animal protein sources meet minimum levels, particularly for larvae and juvenile stages, which have critical requirements for essential amino acids. It is possible to satisfy the nutritional needs of adults with diets containing lower amounts of animal protein and relatively higher levels of plant protein, as the requirement for essential amino acids gradually decreases with age. Common sources of animal protein in aquaculture diets include fish meal, poultry meal, feather meal, and meat by-products. Additionally, it is essential to consider substances that may fully or partly replace organic fertilizers and alternative ingredients as sources of supermolecules in aquafeeds. Thanks to their chemical makeup, Protoctista products may be utilized to enhance the nutritional content of food and animal feed, which are essential for cultivation [10, 11].

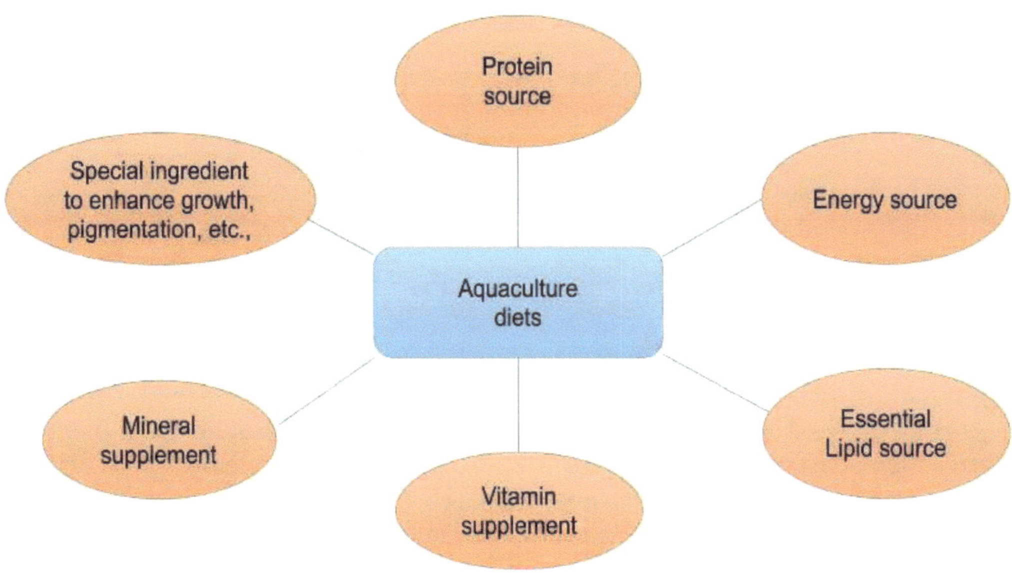

Fig. (1). Essential needs for aquaculture diets.

NUTRITIONAL NECESSITIES

The nutritional balance of feed ingredients affects the utilization of feed and the growth of cultivated species in these areas. Dietary nutrient levels may be influenced by management, environmental conditions, and animal size to achieve optimal performance. Nonetheless, the data on nutritional necessities that are currently available may be used to produce diets for industrial cultivation that are extremely productive and efficient. The nutritional needs of a closely related species whose nutritional needs are known may be utilized when creating a diet for a species whose nutritional needs are unknown. In general, warm- and cold-water species, freshwater species, finfishes, and crustaceans ought to be expected to vary foremost in their nutrient necessities among categories. For a species to develop at its optimum and create a diet, it's crucial to understand the nutritional necessities, particularly for macromolecules, fats, and energy. Inadequate quantities of supermolecules and energy in feed raise production prices and degrade water quality. Supermolecule waste results from diets with meager energy, as a larger share of dietary supermolecules is needed for energy, and the ammonia created will harm the quality of water. To boot, feed and parts ought to provide the specified nutrients in quantities to fulfill the needs of cultivation species [12] (Table **1**).

Table 1. Diet supply necessary for the optimal growth and health of fish (Craig & Helfrich, 2002).

Requirements	Quantity (in %)
Protein	18-50%
Lipid	10-25%
Carbohydrates	15-20%
Ash	<8.5%
Phosphorous	<1.5%
Water	<10%
Vitamins & Minerals	Trace amounts

According to a 2003 FAO study, protein is the primary and most expensive component of feed and serves as the essential building block for feed formulation systems [13]. The FAO proposed the concept of "an ideal protein" as a method to identify the specific amino acids required by different fish species [14]. Until the dietary essential amino acid requirements are established through trial-and-error using test diets, the dietary requirements determined using the ideal protein concept (based on the whole-body amino acid profile) provide a useful guideline for formulating diets for cultured aquaculture species [15]. The essential amino acids—histidine, isoleucine, leucine, lysine, methionine, threonine, tryptophan, and valine—are required for optimal growth and metabolism in all tested fish species [16].

The ten essential amino acids required by shrimp are methionine, valine, threonine, isoleucine, leucine, lysine, histidine, arginine, tryptophan, and phenylalanine, according to various studies [17, 18]. Fats, oils, phospholipids, and sterols are among the organic substances classified as lipids. Fish utilize dietary lipids as a major source of energy, sparing proteins, and providing essential fatty acids necessary for maintaining membrane fluidity and permeability, growth, survival, and various physiological activities. Dietary lipids also influence the taste, texture, and quality of processed foods and fish meat [19]. Additionally, dietary lipids are a concentrated and highly digestible source of energy, providing 8-9 kcal/g, which is approximately twice the energy of either protein or carbohydrate [20]. Although carbohydrates are an inexpensive energy source, aquatic animals such as shrimp and fish have a limited capacity to utilize them [21] due to their restricted ability to metabolize and regulate plasma glucose levels [22]. Fish and shrimp utilize carbohydrates less efficiently than terrestrial animals [23]. Moreover, fish are known to have a limited capacity for carbohydrate digestion and metabolism; therefore, excessive carbohydrate intake can lead to metabolic problems [24]. High levels of carbohydrates can slow growth and are often associated with inefficient feed utilization [25].

FEED ADDITIVES

These are various feed additives that can enhance the growth performance of shrimp and fish. Since feed costs represent a significant portion of production expenses in aquaculture, products that improve feed efficiency are particularly important. Proper use of such products can increase the profitability of aquaculture operations. Each health-related, non-nutritive feed additive can affect an animal's physiology either directly or indirectly. Various factors influence multiple physiological systems, and their effects may interact with one another. The use of natural resources, such as medicinal plants, as feed additives to improve feed efficiency and overall aquaculture productivity is well established. Recently, among other feed additives, medicinal herbs and probiotics have been recommended as viable alternatives to antibiotics in aquaculture diets [26].

Since phytogenics include various types of compounds, they have been classified based on their composition, processing, and biological origin. Spices are herbs with a strong aroma or flavor, often used as supplements in human food. Herbs are leafy flowering plants known for their medicinal properties. Essential oils are aromatic, oily liquids extracted from plant materials such as flowers, leaves, fruits, and roots. Oleoresins are extracts obtained from plant materials using nonaqueous solvents [27]. Numerous aquaculture species depend on binding additives, amino acids and their metabolites, and vitamins as key metabolic regulators for feed intake, nutrient utilization, maintenance, growth, immunity, behavior, larval metamorphosis, reproduction, and resistance to environmental stressors and pathogens [28].

By mitigating the negative effects of stress and promoting growth, nutraceuticals act as buffering agents in biological systems. Supplementation with protein-rich diets has been shown to alleviate a variety of stressors. Vitamins, such as C and E, are commonly added to diets to help shellfish and finfish cope with stress. Nutraceutical supplementation can also reduce the impact of environmental stressors, including temperature fluctuations, salinity changes, and chemical exposure, while promoting growth and modulating non-specific immune responses [29]. The use of organic acids and their salts as feed additives in animal production is widely accepted and considered safe. Studies indicate that organic acids enhance nutrient absorption, growth, feed conversion efficiency, competitive exclusion of pathogens, and the overall health of shrimp and fish [30]. When organic acids are active, they can penetrate the semipermeable membranes of certain microorganisms and disrupt their normal functioning.

AQUAFEED INGREDIENT TESTING

A feed ingredient is typically evaluated using a series of basic chemical tests, known as proximate analysis, to determine its content of moisture, crude protein, fat, crude fiber, ash, and digestible carbohydrates. Testing feed ingredients *in vivo* requires extensive trials with live animals. However, *in vitro* techniques allow for rapid assessment of the test substance's nutritional value as well as the potential adverse effects of antinutritional compounds. A promising new approach is the determination of digestibility using an *in vitro* pH-stat assay. This assay simulates protein digestion by replicating the enzymatic activity of the target animal, providing a rapid and controlled measure of the ingredient's nutritional quality. The cost of digestible energy per unit, which focuses on the development of assays to assess available energy, is considered acceptable because energy represents the primary cost factor in formulating aquaculture diets. It is therefore reasonable to focus this type of investigation on digestible energy, as it is the most straightforward energy parameter to quantify in shrimp and fish feed ingredients.

On the other hand, various *in vitro* assays, such as the enzyme digestibility assay, enzyme assay, and potash solubility test, are used with varying degrees of effectiveness to assess the quality of proteins and ingredients. The enzyme digestibility assay, developed in the early 1950s, is arguably the most popular and rapid *in vitro* method for assessing protein quality.

Recent advancements in methods for assessing digestibility in feeds for terrestrial animals, both *in vivo* and *in vitro*, are also valuable for evaluating aquafeed ingredients [31, 32]. One promising approach is the pHstat *in vitro* technique, which simulates the breakdown of protein by the target animal's enzymes. When proteins from comparable animal or plant sources are tested, a strong correlation is observed between the pHstat *in vitro* results and *in vivo* digestibility values [33]. The enzyme digestibility assay, developed in the early 1950s, remains the fastest and most widely used *in vitro* method for evaluating protein quality.

Examining the components of aquafeed can benefit from recent advancements in methods used to assess the digestibility of feeds for terrestrial animals, both *in vivo* and *in vitro* [31, 32]. One promising approach is the pHstat *in vitro* technique, which evaluates the digestibility of feed and feed ingredients by simulating the breakdown of protein by the target animal's enzymes. When proteins from comparable animal or plant sources are tested, the pHstat results show a strong correlation with *in vivo* digestibility values. Enzymatic activity or the presence of certain antinutritional compounds can render portions of an ingredient's protein structure indigestible. Proteins are cleaved by enzymes at specific sites, but some antinutritional compounds can block these sites, hindering proper digestion and reducing the nutritional value of the ingredient in feed. By potentially limiting enzyme access or interfering at the site of enzymatic action, these compounds slow the rate of protein digestion. Additionally, interactions such as the formation of complexes between starch and lipids can influence the outcomes of digestibility assays. These complexes are believed to reduce the digestibility and utilization of ingested carbohydrates and may persist within the gastrointestinal tracts of various aquatic species.

FORMULATION OF AQUAFEED

When formulating supplemental feeds, ingredients can be selected from a well-established list to create a feed mixture that meets the required crude protein levels. Moreover, feed rations can now be calculated in virtually unlimited combinations thanks to advanced software systems and increasing computational power. The use of computers has gradually expanded over the past several years, reducing the time required for ration formulation calculations [34]. The demand for available feed ingredients can change significantly once the least-cost feed formulation is determined using mathematical models based on current market prices, along with minor variations in relative costs. It is therefore essential to implement feed formulation software approaches in parallel with the expansion of intensive aquaculture production systems in many countries. Creating comprehensive biological process databases could potentially provide a feed company with a competitive edge by utilizing broad analysis [35]. Within the Fifty, applied mathematics was developed to reduce feed prices while considering a set of constraints (requirements and ingredient minima and maxima) [36]. The strategies presently used for feed formulation stay largely unchanged. The sensitivity analysis operation of Excel's problem solver routine is often used to estimate the shadow prices of additional ingredients, in addition to a distinctive ingredient mixture, to match diet parameters at the lowest possible price. The ingredient's "shadow pricing" is the very best price at which it might be employed in the formula. The low price necessary for ingredients is too costly to be included within the formula and can be stated as a "shadow worth." Shadow costs are often used to produce usage curves that relate the value of associated ingredients to the amount that might be employed in a particular feed instruction [37].

AQUAFEED FROM WASTE- CIRCULAR ECONOMY

A circular economy is defined by the use of waste to significantly reduce the total amount of waste generated over time. The term "circular bioeconomy" is more specifically applied to describe the use of waste as an ingredient in aquafeed or as a standalone source of nutrients. While the reuse of nutrients and resources through the integration of different aquatic and terrestrial crops has been practiced since the earliest aquaculture systems, modern production methods—particularly intensive systems—have greatly increased the demand for effective waste management. Incorporating waste into the formulation and production of aquafeeds could enhance their value as part of a holistic approach. It has been suggested that Aquafeed 3.0 may ultimately be driven by a circular bioeconomy [38]. A notable example of this approach is Integrated Aquaculture (IA), particularly Integrated Multitrophic Aquaculture (IMTA). A non-fed plant or animal occupying a distinct trophic level can utilize the waste products of a fed organism as a food source. Since the amount of aquafeed supplied is shared between the two co-cultured species, production efficiency is enhanced when both species are cultivated together [39]. The results of a recent laboratory study, in which satisfactory growth of marine shrimp species was fully maintained through the provision of sea urchin egesta, demonstrate the potential of IMTA systems [40]. The social acceptability and economic viability of commercial-scale IMTA production—aside from invertebrate integration, such as scallop-macroalgae co-culture in Sanggou Bay, China [41]—have yet to be fully established. However, studies of aquaponic recirculating systems based on the IMTA principle suggest that economic viability may be achievable under certain conditions [42].

The cultured organisms develop and utilize rich food sources (bio-flocs) when specific organic matter is supplemented to lake production systems. The amount of aquafeed needed is often reduced while still achieving the mandatory production goals, thereby increasing feed potency. In some management techniques, supplemental feed (corn protein meal pellets), intended as an extra source of nutrients for terrestrial animals, is often used as a nutritionally deficient supplement to cultivation production ponds. The leftover, warmed-over feed pellets quickly break down into minute particles that are then absorbed by the biogeographical region's creatures, which develop and reproduce. A number of these nutrients are directly consumed. To fulfill biological process needs, feed and natural lake assemblages were employed in unison [43]. These last two examples show, however, that an on-the-spot feed is also used at lower levels while still achieving the suitable rate and result.

FISH MEAL AND OIL

Small pelagic marine fish, including menhaden, herring, anchovies, and sardines, are used to produce fish meal. When these small fish are crushed, water and oil are extracted. The remaining solids are cooked and ground into meal for consumption. Animal oil is obtained as a byproduct during fish meal production by separating water from the leftover liquid. The species of fish significantly affects the yield of fish meal and oil [44]. In 2008, an average of 4.56 metric tons of live fish was required to produce one kilogram of fish meal. That same year, a total of 4.82 million tonnes of fish meal was produced, along with 1.02 million tonnes of animal oil as a byproduct. Based on the fish-to-oil ratio of 4.56, approximately 21.5 metric tons of live fish are needed to produce one kilogram of animal oil. Fish processing wastes are also used to produce organ meal and oil suitable for use in animal feed. In 2008, 1.23 million tonnes of organ meal were produced, with around 0.26 million tonnes of organ oil generated as a byproduct, consistent with the organ meal-to-oil ratio observed for live fish.

Fish meal is widely used in animal feed due to its high protein content and as a rich source of minerals such as iron and phosphorus. Its high protein content and favorable amino acid profile make fish meal particularly suitable for aquaculture feeds. Although edible oils are also used in aquaculture diets, some studies have shown that farmed fish fed diets high in edible oils have a lower ratio of omega-3 fatty acids to polyunsaturated fatty acids compared to wild fish [45 - 47]. It should be noted, however, that not all researchers agree with this conclusion [48], as the fatty acid profile of farmed fish depends on the oils included in their feed, which are controlled by manufacturers. Nevertheless, fish oil remains commonly used in aquaculture feeds, as a high omega-3 to polyunsaturated fatty acid ratio is believed to confer health benefits in humans, including reducing the risk of cardiovascular disorders [49].

Future fish meal and fish oil supplies may be a significant barrier to the continued growth of the aquaculture sector. Fish meal and oil were initially used extensively in other animal production systems, creating similar bottlenecks. Alternatives arise once nutrient requirements are established and quantified. Using the global poultry sector as an example, despite growth over approximately five decades, fish meal has been used less overall in poultry [50]. Finding ways to reduce the inclusion of fish meal and oil in aquaculture feeds is currently a major focus. Efforts include the use of plant meals and oils, animal processing byproducts, and fish organ meal and oil. Additionally, research is underway to develop new sources of omega-3 fatty acids for feed, including algae rich in DHA and EPA and genetically engineered oilseed crops [49, 51]. Additionally, considerable research has focused on feeding strategies that reduce the FCR, thereby lowering the amount of fish meal and fish oil required per unit of production. Efforts to develop improved feed and feeding strategies have been successful. Between 1995 and 2007, the average FCR for major species decreased from 1.95 to 1.75, fish meal inclusion declined from 25.5% to 14%, and fish oil inclusion dropped from 7.5% to 4.4%, according to Naylor [52]. Despite these encouraging results, the total amount of fish meal and oil used in aquaculture feeds increased over the study period due to the significant expansion of aquaculture supported by feed. A total of 23.2 million tonnes of fish were caught in 2011 for non-food purposes, primarily for the production of fish meal and fish oil. The large quantities of marine fish used in aquaculture, compared to 63.6 million tonnes of total aquaculture production in 2011, do not necessarily indicate inefficient use of fish meal and fish oil in marine fisheries. However, NGOs argue that aquaculture's substantial demand for fish meal and fish oil in feed could have the most significant negative impacts on resource use, the environment, and society [53]. Alder reported that since 1961, 10-20% of marine fisheries landings were not used for human consumption, supporting the idea that fish used to produce fish meal and oil are not intended for direct human use [54]. The resources for fish meal and oil are at risk of overexploitation, a reality acknowledged by those in the aquaculture sector. Because many species of larger carnivorous fish rely on small marine fish for sustenance, this could lead to a shortage of fish meal and oil and negatively impact marine ecosystems. Fish meal and fish oil production is part of global capture fisheries, and to calculate total world fisheries production, capture fisheries output is added to global aquaculture production. Naylor asserts that aquaculture reduces global fish productivity if the amount of fish meal and oil used in aquaculture feed requires more live fish than the quantity of aquatic animals produced from that feed [52]. NGOs often use the Fish-In-Fish-Out (FIFO) ratio as an indicator of sustainable aquaculture practices. A FIFO ratio greater than 1.0 indicates that more wild fish are used to produce the feed than the amount of farmed fish produced in a given year. NGOs argue that FIFO ratios in the aquaculture industry should be minimized and are particularly important for aquaculture species or facilities with a FIFO ratio exceeding 1.0. The typical FIFO ratio for salmon, trout, eel, marine fish, and shrimp decreased from 4.7 in 1995 to 3.1 in 2006, demonstrating the effectiveness of this strategy. Several important aquaculture species typically have FIFO ratios below 1.0, including freshwater shrimp (0.6), tilapia (0.4), milkfish (0.2), and Chinese carp (0.4) [55].

An estimated 19.7 million tonnes of feed-based aquaculture production could be generated. This figure was calculated by dividing the 34.4 million tonnes of aquaculture feed produced globally in 2012 by the average FCR of 1.75. Aquaculture feeds contain 81% fish oil and 63% fishmeal. The FIFO ratio should be based on fish oil, as it is a byproduct of fishmeal production, and its yield is lower than that of fishmeal. In 2012, 1.02 million tonnes of fish oil were produced, of which 0.83 million tonnes were used in aquaculture. Approximately 17.8 million tonnes of wild fish were required to produce the fish oil for aquaculture feed, assuming a conversion rate of 21.5 tonnes of wild fish per tonne of fish oil. For feed-based aquaculture, the FIFO ratio is therefore estimated at 0.90. Furthermore, fisheries products not typically used for human consumption are transformed into edible aquaculture products, which can increase global food output. It should be noted that this outcome exceeds what might be inferred from the FIFO ratio, even though aquaculture consumes the majority of fishmeal and fish oil globally [52, 55].

ZOOPLANKTON AS FEED FOR FISHES

In regions where available, the world's oceans contain immense stocks of herbivorous copepods and malacostracan crustaceans that are abundant and potentially exploitable. The Southern Ocean's Elephant Island region had an average Antarctic malacostracan crustacean biomass of 1.5 to 31 g wt m^{-2} from 1977 to 1997 [56]. Since the standing stock biomass of Antarctic malacostracan crustaceans is estimated to be 500 million t [57], the annual production could be at least 100 million t/yr (assuming a lifespan of six years). Based on estimates of 74 million t/yr of red feed *Calanus finmarchicus* produced annually in the Nordic Seas [58], further research is needed to determine whether this species is suitable as a feed resource. Red feed is currently harvested commercially only to a limited extent, and this collection is primarily part of focused research projects. According to studies, the growth rate of Calanus species is inversely proportional to biomass, which is a general biological principle. Simulations using realistic 3D biological models have shown that, even under 90% simulated harvesting effort, increased food availability supports rapid recovery of the *C. finmarchicus* stock. It is evident that the annual production of *Antarctic malacostracan* crustaceans and red feedstock is highly valued by the world's fish and squid fisheries. A marine resource relies on the catch of oceanic forage fish, which accounts for approximately 10-20% of total production. Moreover, the higher-trophic-level fish would move one trophic level lower in the food web if terrestrial animal products replace fish products in the feed of carnivorous fish [59].

There is a lack of data concerning Antarctic malacostracan crustaceans, although significant concerns exist regarding the use of red feed and Antarctic malacostracan crustaceans as food for fish, birds, and marine mammals in the Norwegian Sea and Antarctic Ocean, respectively [54, 60]. Because malacostracan crustaceans, and potentially red feed, are vital for fisheries as well as a food source for birds and mammals, this issue has management and political implications that require careful study [60]. Since these animal resources are finite and harvested rather than naturally replenished, they cannot fully support farmed carnivorous fish to the trophic level desired. However, if these resources are used with minimal negative effects on humans and marine ecosystems, they can provide essential components, such as Highly Unsaturated Fatty Acids (HUFA), necessary for the development of mariculture in the near future.

AGRICULTURE PLANTS FEED

Freshwater aquaculture largely utilizes a wide variety of plant resources or agricultural waste, with only trace amounts of fish meal [55]. As noted earlier, significant research initiatives are beginning to promote the use of higher plant proteins and oils in pellet feed for aquaculture. For example, corn, soy flour, protein meal, gluten, peas, and oilseed meal are important substitutes for fish meal [52, 61]. Oils such as soybean oil, flaxseed oil, and other plant oils are used to replace fish oil [61, 62]. This approach has been effective because it has allowed aquaculture productivity to increase despite the stagnating supply from marine feed providers. However, plant oils cannot satisfy the HUFA requirements of carnivorous species because they contain only short-chain C18 n-3 and n-6 fatty acids (ALA and LA). Consequently, for certain applications, the progressive inclusion of higher plant oils in carnivore feeds is constrained by this physiological limitation, creating a tipping point where animal health, welfare, and growth can be negatively affected [63].

HIGHER GENETICALLY CHANGED (GMO) PLANTS

Over the past decade, businesses and research institutions have invested heavily in the development of transgenic higher plants (GMOs) that will produce HUFAs, along with DHA, EPA, and ARA [62, 64 - 66]. This indicates a vital improvement in independent agency production among the concerned analysis teams (up to 26% of total fatty acids). However, DHA yields are still relatively low (up to 3%), making it challenging to assess the current state of this effort [65]. Utilizing the existing biological science strategies, progress has been made quickly. A pronto out there, DHA supply derived from higher GMO plants might essentially alter the outlook for the expansion of global cultivation and animal production.

MARINE MACROALGAE AS AQUAFEED

Marine macroalgae were cultivated at a rate of approximately 15 million t yr^{-1} in 2007. Additionally, 1.0 million t of natural resources were extracted. Harvested macroalgae have traditionally been used to produce marine biopolymers, which are widely applied as food additives, for pharmacological and medical purposes, and for direct human consumption in Asian countries [67]. Macroalgae have long been used in mollusc-based mariculture, and attention has recently turned to them as a source of bioenergy [68, 69]. Marine macroalgae, like higher plants, contain HUFA [70 - 72]. Despite their relatively modest protein content, macroalgae represent a nearly untapped resource that could replace marine fish oils in the diets of marine carnivores. The protein content of macroalgae varies by species and physiological status [73]. They must be thoroughly processed before use in fish feed, and further progress is still required in this area.

MICROALGAE AS FEED

Since several species of microalgae have a high protein content and their lipids may contain high levels of HUFA, they appear nutritionally suitable as a resource for mariculture feed [72, 74]. Additionally, cultured microalgae play a crucial role in the on-site production of juvenile marine fish, shrimp, and molluscs [67, 75]. However, due to their reliance on light as an energy source for biosynthesis, production costs are comparatively high. According to Walker, protists are inefficient at converting light into biomass, and increasing light intensity can lead to photoinhibition, ultimately damaging the photosystems [76, 77]. Compared to algal cultures grown mixotrophically or heterotrophically on organic substrates, the biomass yields in photobioreactors are lower [78].

The use of microalgal lipids as highly valuable DHA and HUFA components in aquaculture feed is feasible, but microalgae are unlikely to replace other essential bulk feed sources. Additionally, transgenic higher plants, macroalgae, or yeast can be developed using genetic material from microalgae. Thraustochytrids, which are fungus-like protists, constitute a taxonomically distinct group of DHA-rich microorganisms that can be cultivated in bioreactors [72]. Some strains are capable of producing substantial biomass, and in the future, oils from thraustochytrids may provide a high-quality alternative source of DHA for mariculture.

BIOMASS OF MICROORGANISMS

Yeast or microorganism biomass is an alternative double feed source in marine agriculture. The feeding trials on rainbow trout *Oncorhynchus mykiss*, employing a feed containing up to 27% microorganism supermolecule meal, have generated favourable results. Microorganism-derived supermolecule meal may be a new resource created to support alkanes from fossil fuels [79]. Carnivorous warm-blooded animals have also found the feed to be palatable [80]. HUFA was not present in the microorganisms used. The main feed ingredient for mariculture can be an inexpensive, nutrient-dense SCB with HUFA, cultivated using a low-cost carbon waste source, such as alkanes. Although the use of alkanes is not mandatory, this concept for feed production is particularly attractive because it allows food production based on waste materials, sewage, and fossil fuels, which can close the nutrient cycle. The organism is the crucial component; currently, no suitable organism has yet been identified and farmed on a large scale, although it is being sought [81]. Presumably, suitable organisms are being developed using modern genetic techniques and may be transgenic.

A RANGE OF FISH FOODS

Farm-made single-ingredient feeds, such as rice bran or refined feed, are examples of feeds used as on-farm feeds. Another type of farm-made feed that can be affordably prepared for a small-scale culture project is kitchen waste. Farm-made or on-farm feeds are prepared from one or more natural and/or artificial ingredients and are produced solely for use in aquaculture operations, not for commercial profit. These feeds are usually mixtures of ingredients that have undergone some form of processing (basic mixing, grinding, or cooking) performed on the farm or in small processing units. They are often used in small-scale semi-intensive culture systems.

Commercial feeds, on the other hand, are designed and produced using a precise combination of multiple ingredients in specific proportions to meet defined quality targets in terms of size, texture, stability, and nutritional composition at a reasonable cost. Most intensive and semi-intensive practices use commercial complete feeds. Some feeds are formulated with antibiotics to treat bacterial diseases such as vibriosis and are often labeled as medicated feeds.

TYPES OF FEED

Feeds will be created through extrusion, which results in enlarged, floating, or buoyant feeds, or by a steam process that creates compacted, pressure-pelleted (sinking) feeds.

Floating Feeds

Generally speaking, floating feeds have several advantages over their sinking counterparts. Screws on the barrel of the extruder machine propel the raw materials, which are then cooked for around thirty seconds at 120-175 °C. Under the air mass, the uniformly cooked mixture is pressed through a die. The pressure differential causes the material to expand. This preparation method makes floating feeds more digestible, and the heat and pressure also inactivate harmful enzymes. Increased starch gelatinization slows disintegration, making the feed more stable in water and giving the fish longer to consume it completely. Additionally, the farmer can observe the feeding behavior of the fish directly and adjust the feeding rate accordingly. Determining whether the feeding rate is too high or too low is crucial for optimal fish growth and feed utilization. Another benefit is that farmers can monitor the health status of the fish as they surface to feed. On the other hand, floating feed can be disadvantageous in terms of competition and for some fish species' feeding behavior.

Sinking feeds

Solid feed pellets, called sinking feeds, are submerged during application. For example, bottom-feeding shrimp prefer sinking pellets (density greater than water, 1 g/cm^3) and will not consume floating feed [82]. When it comes to biomass in the pond, the farmer often underestimates feeding rates, either underfeeding or overfeeding, compared to the actual requirement. Underfeeding reduces weight gain at a given age and extends the culture period, while overfeeding leads to feed wastage and deteriorates water quality, which can cause various problems.

PRESERVING FEED QUALITY

The purity and quality of aquafeeds are influenced by a variety of factors. The origin, processing, handling, and storage of feed materials, as well as other market-related factors, can affect feed standards and safety at multiple levels [83]. Using high-quality ingredients helps establish feed quality, and the sourcing of raw materials should follow appropriate criteria for quality, traceability, environmental impact, and safety.

The quality of feed as it leaves the manufacturer is generally as intended; however, by the time it reaches the farmer's pond, that quality may have deteriorated. Large farms typically receive commercial feed in truckloads of bulk feed, which is often stored outdoors in bins [82]. Once stored, finished feeds can undergo deteriorative changes that reduce their nutritional value below minimum requirements and impair their flavor and appearance. To minimize microclimatic effects and maintain quality, feeds should be stored in dry, well-ventilated warehouses away from direct sunlight. All feed should be used within two months of production and inspected regularly. Prolonged storage can lead to mold growth, nutrient loss, and fat rancidity. Excessive handling of feed bags can damage them and create wasteful dust that fish do not consume. To prevent contamination, pests such as mice, rats, and cockroaches should be strictly controlled in storage areas. Correct storage is a simple but crucial step in maintaining feed quality. Recognizing that animal feed safety is essential for human food safety, nutritionists, industry personnel, farmers, and other relevant authorities should focus on proper feed manufacturing procedures [83].

STORAGE

The storage duration of feed should be managed based on how it will be used in aquaculture operations. Depending on the location of the mill, the quantity of feed, and the proximity of the production facility, feed is typically used within 4-6 months in grow-out operations with rapid turnover. In other cases, it may be stored for 4-6 months. Therefore, specific precautions should be taken when feed is stored for extended periods. Both wet and dry feeds are available in the aquaculture industry, although for reasons of cost and convenience, dry feeds are currently the most common. Examples include extruded feeds, hard pellets, crumbles, and flakes. Dry feed should be stored in a dry, well-ventilated area that protects against sudden temperature fluctuations. Moisture content and ambient temperature are the main factors affecting feed storage. Cooler interior temperatures are preferred as they help prevent mold growth and insect activity (in a warehouse, silo, or bin) [84]. However, minimizing extreme temperature variations is even more important than maintaining a specific temperature. Temperature differences caused by sunlight on the storage structure and day-night cycles can lead to condensation, resulting in water dripping into the feed. This moisture accelerates the growth of mold already present on the feed surface. Therefore, a proper storage facility should be designed to reduce condensation and provide adequate protection against pests. It is important to understand that the feed's manufacturing process directly affects its quality, particularly its moisture content. For faster processing and improved end-use quality, water in the form of steam is added to the feed mash in the conditioner before pelleting. During cooling, the latent heat removal should be used to remove 3-6% of the moisture [85]. If this moisture removal is insufficient, the feed will have higher moisture content, which can lead to rapid mold growth. To allow the finished pellet to stabilize with environmental conditions and prevent heat transfer from the pellet to the storage structure—which could cause problems—the final pellet temperature should be cooled to 5-8°C above the ambient temperature. As part of an integrated facility, the feed producer must follow all standard procedures for safe storage of ingredients. These procedures include aeration, inventory rotation, fumigation, and the use of mold inhibitors and grain protectants. Therefore, it is essential to monitor and understand feed quality upon delivery to the farm before storage. Losses caused by improper storage, mold growth, pest infestation, nutrient degradation, or fat rancidity can be significantly reduced with knowledge of the key factors contributing to feed deterioration and attention to proper storage conditions.

CONCLUSION

Feed-based aquaculture appears to be a relatively resource-efficient approach for increasing the supply of animal protein for human consumption. About 81% of animal oil is used in aquaculture diets, compared to only 63% of fish meal. Inclusion rates of fish meal and oil in aquaculture feeds have decreased significantly due to extensive research on alternatives. The large quantity of small pelagic fish required to produce fish meal and oil represents the main environmental impact of feed-based aquaculture. NGOs argue that overfishing of pelagic fisheries could substantially alter marine food chains and reduce populations of species that feed on these fish. To enhance the nutritional content or utilization of feed ingredients or to modify nutrient supply in current commercial feeds, a variety of novel technologies are being developed and tested. These include enzymes, genetically engineered feed components, and feed-processing technologies designed to increase nutrient availability, meet the specific requirements of farmed species, and reduce nutrient excretion. By providing nutrients in the correct proportions, these specialized feeds and technologies support "precision feeding" of aquaculture species.

A deeper understanding of marine ecosystems can make it possible to define the "natural" baseline of their biological functions, monitor changes over time, and predict their potential impacts on these systems. The growth and sustainability of a robust marine biotechnology sector require a strong property rights framework, verified governmental infrastructure, public and private research and development funding, support for early-stage start-ups through working capital, public acceptance *via* awareness and engagement, and a reliable pipeline of skilled human resources.

REFERENCES

[1] Tacon AG, Hasan MR, Metian M. Demand and supply of feed ingredients for farmed fish and crustaceans: trends and prospects. FAO Fish Aquac Tech Pap 2011; (564): I

[2] Boyd CE, Queiroz J, McNevin A. Perspectives on the responsible aquaculture movement. World Aquac 2013; 44(4): 14-21.

[3] Boyd CE, Tucker CS. Pond Aquaculture Water Quality Management Kluwer Academic Publishers. 1998.
[http://dx.doi.org/10.1007/978-1-4615-5407-3]

[4] Alday-Sanz V, editor. The shrimp book. Nottingham: Nottingham University Press; 2010.
[http://dx.doi.org/10.1079/9781800629349.0000]

[5] Boyd CE, Li L. Environmental issues in pond fertilization. Aquaculture Pond Fertilization. 2012 Apr 12: 65-72.
[http://dx.doi.org/10.1002/9781118329443.ch5]

[6] Moran CA, Connolly A. The global feed industry. AFMA Matrix 2016; 25(2): 10-1.

[7] Glencross BD. Exploring the nutritional demand for essential fatty acids by aquaculture species. Rev Aquacult 2009; 1(2): 71-124.
[http://dx.doi.org/10.1111/j.1753-5131.2009.01006.x]

[8] Hardy RW. New developments in aquatic feed ingredients, and potential of enzyme supplements. Avances en Nutrición Acuicola. 2000.

[9] Boyd, C.E., Tucker, C.S., McNevin, A., Bostick, K., Clay, J. Indicators of resource use efficiency and environmental performance in fish and crustacean aquaculture. Rev. Fish. Sci. 2007; 15, 327e360.
[http://dx.doi.org/10.1080/10641260701624177]

[10] Dewi AP, Nursyam H, Hariati AM. Response of fermented Cladophora containing diet on growth performances and feed efficiency of Tilapia (Oreochromis sp.). Int J Agron Agric Res 2014; 5(6): 78-85.

[11] Jamali H, Ahmadifard N, Abdollahi D. Evaluation of growth, survival and body composition of larval white shrimp (Litopenaeus vannamei) fed the combination of three types of algae. Int Aquatic Research 2015; 7(2): 115-22.
[http://dx.doi.org/10.1007/s40071-015-0095-9]

[12] Tacon AG. The nutrition and feeding of farmed fish and shrimp: a training manual. Vol. 2, Nutrient sources and composition. Rome: FAO; 1990.

[13] Shiau SY. Nutrient requirements of penaeid shrimps. Aquaculture 1998; 164(1-4): 77-93.
[http://dx.doi.org/10.1016/S0044-8486(98)00178-1]

[14] Wilson RP. Protein and amino acid requirements of fishes. 1989.

[15] Wilson P. Amino acid nutrition of fish: a new method of estimating requirement values. InProceedings of the 20th US-Japan Symposium of Aquaculture Nutrition 1993 (pp. 49-54). NOAA, Department of Commerce.

[16] Lim C, Akiyama DM. Nutrient requirements of penaeid shrimp. Nutrition and utilization technology in aquaculture. 1995: 60-73.

[17] Fox JM, Davis DA, Wilson M, Lawrence AL. Current status of amino acid requirement research with marine penaeid shrimp. Avances en Nutrición Acuícola 2006.

[18] Stickney RR, Hardy RW. Lipid requirements of some warmwater species. Aquaculture 1989; 79(1-4): 145-56.
[http://dx.doi.org/10.1016/0044-8486(89)90455-9]

[19] Chuang JL. Nutrient requirements, feeding and culturing practices of Penaeus monodon: a review The Nutrition of Prawns. Basel: F. Hoffman-La Roche Ltd. 1990.

[20] Shiau SY. Utilization of carbohydrates in warmwater fish—with particular reference to tilapia, Oreochromis niloticus X O. aureus. Aquaculture 1997; 151(1-4): 79-96.
[http://dx.doi.org/10.1016/S0044-8486(96)01491-3]

[21] Zainuddin H, Aslamyah S. Effect of dietary carbohydrate levels and feeding frequencies on growth and carbohydrate digestibility by white shrimp Litopenaeus vannamei under laboratory conditions. J Aquac Res Dev 2014; 5(6): 274.

[22] Mohapatra M, Sahu NP, Chaudhari A. Utilization of gelatinized carbohydrate in diets of *Labeo rohita* fry. Aquacult Nutr 2003; 9(3): 189-96.
[http://dx.doi.org/10.1046/j.1365-2095.2003.00243.x]

[23] Hemre G-I, Mommsen TP, Krogdahl Å. Carbohydrates in fish nutrition: effects on growth, glucose metabolism and hepatic enzymes. Aquacult Nutr 2002; 8(3): 175-94.
[http://dx.doi.org/10.1046/j.1365-2095.2002.00200.x]

[24] Zhou C, Ge X, Liu B, Xie J, Chen R, Ren M. Effect of high dietary carbohydrate on the growth

performance, blood chemistry, hepatic enzyme activities and growth hormone gene expression of Wuchang Bream (Megalobrama amblycephala) at two temperatures. Asian-Australas J Anim Sci 2015; 28(2): 207-14.
[http://dx.doi.org/10.5713/ajas.13.0705] [PMID: 25557816]

[25] Dada A, Olugbemi BD. Dietary effects of two commercial feed additives on growth performance and body composition of African catfish Clarias gariepinus fingerlings. Afr J Food Sci 2013; 7(9): 325-8.
[http://dx.doi.org/10.5897/AJFS2013.1029]

[26] Jacela J, DeRouchey J, Tokach M, *et al.* Feed additives for swine: Fact sheets – prebiotics and probiotics, and phytogenics. J Swine Health Prod 2010; 18(3): 132-6.
[http://dx.doi.org/10.54846/jshap/624e]

[27] Tincy V, Mishal P, Akhtar MS, Pal AK. Turning Challenges into Opportunities. World Aquac 2014; (Jun): 67.

[28] Manush SM, Pal AK, Das T, Mukherjee SC. Dietary high protein and vitamin C mitigate stress due to chelate claw ablation in Macrobrachium rosenbergii males. Comp Biochem Physiol A Mol Integr Physiol 2005; 142(1): 10-8.
[http://dx.doi.org/10.1016/j.cbpb.2005.07.004] [PMID: 16109496]

[29] Romano N, Koh CB, Ng WK. Dietary microencapsulated organic acids blend enhances growth, phosphorus utilization, immune response, hepatopancreatic integrity and resistance against Vibrio harveyi in white shrimp, Litopenaeus vannamei. Aquaculture 2015; 435: 228-36.
[http://dx.doi.org/10.1016/j.aquaculture.2014.09.037]

[30] Cruz-Suárez LE, Nieto-López M, Guajardo-Barbosa C, Tapia-Salazar M, Scholz U, Ricque-Marie D. Replacement of fish meal with poultry by-product meal in practical diets for Litopenaeus vannamei, and digestibility of the tested ingredients and diets. Aquaculture 2007; 272(1-4): 466-76.
[http://dx.doi.org/10.1016/j.aquaculture.2007.04.084]

[31] Lemos D, Lawrence AL, Siccardi AJ III. Prediction of apparent protein digestibility of ingredients and diets by *in vitro* pH-stat degree of protein hydrolysis with species-specific enzymes for juvenile Pacific white shrimp Litopenaeus vannamei. Aquaculture 2009; 295(1-2): 89-98.
[http://dx.doi.org/10.1016/j.aquaculture.2009.06.011]

[32] Lemos D, Nunes AJP. Prediction of culture performance of juvenile Litopenaeus vannamei by *in vitro* (pH-stat) degree of feed protein hydrolysis with species-specific enzymes. Aquacult Nutr 2008; 14(2): 181-91.
[http://dx.doi.org/10.1111/j.1365-2095.2007.00536.x]

[33] Al-Deseit B. Least-cost broiler ration formulation using linear programming technique. J Anim Vet Adv 2009; 8(7): 1274-8.

[34] Ghosh D, Sathianandan TV, Vijayagopal P. Feed formulation using linear programming for fry of catfish, milkfish, tilapia, Asian sea bass, and grouper in India. J Appl Aquacult 2011; 23(1): 85-101.
[http://dx.doi.org/10.1080/10454438.2011.549781]

[35] Baum EL, Fletcher HB, Walkup HG, Stadelman WJ. An application of profit maximizing techniques to experimental fryer input-output data. Poult Sci 1953; 32(3): 375-83.
[http://dx.doi.org/10.3382/ps.0320375]

[36] Udo IU, Ndome CB, Asuquo PE. Use of stochastic programming in least-cost feed formulation for african catfish (Clarias gariepinus) in semi-intensive culture system in Nigeria. J Fish Aquat Sci 2011; 6(4): 447-55.
[http://dx.doi.org/10.3923/jfas.2011.447.455]

[37] Colombo SM, Turchini GM. 'Aquafeed 3.0': creating a more resilient aquaculture industry with a circular bioeconomy framework. Rev Aquacult 2021; 13(3): 1156-8.
[http://dx.doi.org/10.1111/raq.12567]

[38] Boyd CE, D'Abramo LR, Glencross BD, *et al.* Achieving sustainable aquaculture: Historical and

current perspectives and future needs and challenges. J World Aquacult Soc 2020; 51(3): 578-633. [http://dx.doi.org/10.1111/jwas.12714]

[39] Jensen S, Ragnarsdottir O, Johannsson R. Marine sources of furan fatty acids. J Aquat Food Prod Technol 2019; 28(1): 74-83. [http://dx.doi.org/10.1080/10498850.2018.1561569]

[40] Shi H, Zheng W, Zhang X, Zhu M, Ding D. Ecological–economic assessment of monoculture and integrated multi-trophic aquaculture in Sanggou Bay of China. Aquaculture 2013; 410-411: 172-8. [http://dx.doi.org/10.1016/j.aquaculture.2013.06.033]

[41] Greenfeld A, Becker N, McIlwain J, Fotedar R, Bornman JF. Economically viable aquaponics? Identifying the gap between potential and current uncertainties. Rev Aquacult 2019; 11(3): 848-62. [http://dx.doi.org/10.1111/raq.12269]

[42] D'Abramo LR, New MB. Nutrition, feeds and feeding. Freshwater prawns: biology and farming. 2010: 218-38.

[43] Shepherd CJ, Pike IH, Barlow SM. Sustainable feed resources of marine origin. European Aquaculture Society Special Publication 2005; 35: 59-66.

[44] Bell JG, McEvoy J, Tocher DR, McGhee F, Sargent JR, Campbell PJ. Replacement of fish oil with rapeseed oil in diets of Atlantic salmon (Salmo salar) affects tissue lipid compositions and hepatocyte fatty acid metabolism. J Nutr 2001; 131(5): 1535-43. [http://dx.doi.org/10.1093/jn/131.5.1535] [PMID: 11340112]

[45] Alasalvar C, Taylor KDA, Zubcov E, Shahidi F, Alexis M. Differentiation of cultured and wild sea bass (Dicentrarchus labrax): total lipid content, fatty acid and trace mineral composition. Food Chem 2002; 79(2): 145-50. [http://dx.doi.org/10.1016/S0308-8146(02)00122-X]

[46] Lenas D, Nathanailides C. High content of n-6 fatty acids in the flesh of farmed fish sea bass Dicentrarchus labrax L. and gilthead sea bream Sparus aurata L.: Implications for the human health. Int Aquatic Research 2011; 3(3): 181.

[47] Hardy RW. Farmed fish and omega-3 fatty acids. Aquacult Mag 2003; 29(2): 63-5.

[48] Adarme-Vega TC, Lim DKY, Timmins M, Vernen F, Li Y, Schenk PM. Microalgal biofactories: a promising approach towards sustainable omega-3 fatty acid production. Microb Cell Fact 2012; 11(1): 96. [http://dx.doi.org/10.1186/1475-2859-11-96] [PMID: 22830315]

[49] Delgado CL. Fish to 2020: Supply and demand in changing global markets. World Fish 2003.

[50] Miller MR, Nichols PD, Carter CG. *n* -3 Oil sources for use in aquaculture – alternatives to the unsustainable harvest of wild fish. Nutr Res Rev 2008; 21(2): 85-96. [http://dx.doi.org/10.1017/S0954422408102414] [PMID: 19087364]

[51] Naylor RL, Hardy RW, Bureau DP, *et al.* Feeding aquaculture in an era of finite resources. Proc Natl Acad Sci USA 2009; 106(36): 15103-10. [http://dx.doi.org/10.1073/pnas.0905235106] [PMID: 19805247]

[52] Boyd C, McNevin A. Aquaculture, resource use, and the environment. John Wiley & Sons; 2015 Feb 23. [http://dx.doi.org/10.1002/9781118857915]

[53] Alder J, Campbell B, Karpouzi V, Kaschner K, Pauly D. Forage fish: from ecosystems to markets. Annu Rev Environ Resour 2008; 33(1): 153-66. [http://dx.doi.org/10.1146/annurev.environ.33.020807.143204]

[54] Tacon AGJ, Metian M. Global overview on the use of fish meal and fish oil in industrially compounded aquafeeds: Trends and future prospects. Aquaculture 2008; 285(1-4): 146-58. [http://dx.doi.org/10.1016/j.aquaculture.2008.08.015]

[55] Siegel V, Loeb V, Gröger J. Krill (Euphausia superba) density, proportional and absolute recruitment and biomass in the Elephant Island region (Antarctic Peninsula) during the period 1977 to 1997. Polar Biol 1998; 19(6): 393-8.
[http://dx.doi.org/10.1007/s003000050264]

[56] Nicol S, Endo Y. Fisheries technical paper 367: Krill fisheries of the world. Food and Agriculture Organization 1997; 13: 56-70.

[57] Aksnes DL, Blindheim J. Circulation patterns in the North Atlantic and possible impact on population dynamics of *Calanus finmarchicus*. Ophelia 1996; 44(1-3): 7-28.
[http://dx.doi.org/10.1080/00785326.1995.10429836]

[58] Olsen Y. MARICULT Research Programme: background, status and main conclusions InSustainable Increase of Marine Harvesting: Fundamental Mechanisms and New Concepts. Dordrecht: Springer 2002; pp. 1-10.

[59] Tiller RG. New resources in old waters: The potential of national and international conflicts deriving from the future harvest of Calanus finmarchicus.

[60] Gatlin DM III, Barrows FT, Brown P, *et al.* Expanding the utilization of sustainable plant products in aquafeeds: a review. Aquacult Res 2007; 38(6): 551-79.
[http://dx.doi.org/10.1111/j.1365-2109.2007.01704.x]

[61] Opsahl-Ferstad HG, Rudi H, Ruyter B, Refstie S. Biotechnological approaches to modify rapeseed oil composition for applications in aquaculture. Plant Sci 2003; 165(2): 349-57.
[http://dx.doi.org/10.1016/S0168-9452(03)00194-8]

[62] Tocher DR. Fatty acid requirements in ontogeny of marine and freshwater fish. Aquacult Res 2010; 41(5): 717-32.
[http://dx.doi.org/10.1111/j.1365-2109.2008.02150.x]

[63] Kim SH, Kim SY, Kim JB, Roh KH, Kim YM, Park JS. Biosynthesis of polyunsaturated fatty acids: metabolic engineering in plants. J Appl Biol Chem 2009; 52(3): 93-102.
[http://dx.doi.org/10.3839/jabc.2009.017]

[64] Nichols PD, Petrie J, Singh S. Long-chain omega-3 oils-an update on sustainable sources. Nutrients 2010; 2(6): 572-85.
[http://dx.doi.org/10.3390/nu2060572] [PMID: 22254042]

[65] Venegas-Calerón M, Sayanova O, Napier JA. An alternative to fish oils: Metabolic engineering of oil-seed crops to produce omega-3 long chain polyunsaturated fatty acids. Prog Lipid Res 2010; 49(2): 108-19.
[http://dx.doi.org/10.1016/j.plipres.2009.10.001] [PMID: 19857520]

[66] Wikfors GH, Ohno M. Impact of algal research in aquaculture. J Phycol 2001; 37(6): 968-74.
[http://dx.doi.org/10.1046/j.1529-8817.2001.01136.x]

[67] Gao K, McKinley KR. Use of macroalgae for marine biomass production and CO2 remediation: a review. J Appl Phycol 1994; 6(1): 45-60.
[http://dx.doi.org/10.1007/BF02185904]

[68] Robertson-Andersson DV, Potgieter M, Hansen J, Bolton JJ, Troell M, Anderson RJ, Halling C, Probyn T. Integrated seaweed cultivation on an abalone farm in South Africa. InNineteenth International Seaweed Symposium 2007 (pp. 129-145). Springer, Dordrecht.

[69] Radwan S. Sources of C20-polyunsaturated fatty acids for biotechnological use. Appl Microbiol Biotechnol 1991; 35(4): 421-30.
[http://dx.doi.org/10.1007/BF00169743]

[70] Burtin P. Nutritional value of seaweeds. Electronic journal of Environmental, Agricultural and Food chemistry. 2003; 2(4): 498-503.

[71] Guschina IA, Harwood JL. Lipids and lipid metabolism in eukaryotic algae. Prog Lipid Res 2006;

45(2): 160-86.
[http://dx.doi.org/10.1016/j.plipres.2006.01.001] [PMID: 16492482]

[72] Fleurence J. Seaweed proteins. Trends Food Sci Technol 1999; 10(1): 25-8.
[http://dx.doi.org/10.1016/S0924-2244(99)00015-1]

[73] González López CV, García MCC, Fernández FGA, Bustos CS, Chisti Y, Sevilla JMF. Protein measurements of microalgal and cyanobacterial biomass. Bioresour Technol 2010; 101(19): 7587-91.
[http://dx.doi.org/10.1016/j.biortech.2010.04.077] [PMID: 20494571]

[74] Reitan KI, Rainuzzo JR, Øie G, Olsen Y. A review of the nutritional effects of algae in marine fish larvae. Aquaculture 1997; 155(1-4): 207-21.
[http://dx.doi.org/10.1016/S0044-8486(97)00118-X]

[75] Falkowski PG, Raven JA. Aquatic photosynthesis Princeton University Press. Princeton, New Jersey 2007.
[http://dx.doi.org/10.1515/9781400849727]

[76] Walker DA. Biofuels, facts, fantasy, and feasibility. J Appl Phycol 2009; 21(5): 509-17.
[http://dx.doi.org/10.1007/s10811-009-9446-5]

[77] Sijtsma L, de Swaaf ME. Biotechnological production and applications of the ω-3 polyunsaturated fatty acid docosahexaenoic acid. Appl Microbiol Biotechnol 2004; 64(2): 146-53.
[http://dx.doi.org/10.1007/s00253-003-1525-y] [PMID: 14740186]

[78] Aas TS, Hatlen B, Grisdale-Helland B, Terjesen BF, Bakke-McKellep AM, Helland SJ. Effects of diets containing a bacterial protein meal on growth and feed utilisation in rainbow trout (Oncorhynchus mykiss). Aquaculture. 2006;261(1):357–368.
[http://dx.doi.org/10.1016/j.aquaculture.2006.07.033]

[79] Skrede A, Ahlstrøm Ø. Bacterial protein produced on natural gas: a new potential feed ingredient for dogs evaluated using the blue fox as a model. J Nutr 2002; 132(6) (Suppl. 2): 1668S-9S.
[http://dx.doi.org/10.1093/jn/132.6.1668S] [PMID: 12042487]

[80] Hinzpeter I, Shene C, Masson L. Biotechnological alternatives for omega-3 polyunsaturated fatty acids production. Grasas Aceites 2006; 57(3): 336-42.
[http://dx.doi.org/10.3989/gya.2006.v57.i3.57]

[81] Craig SR, Helfrich LA, Kuhn D, Schwarz MH. Understanding fish nutrition, feeds, and feeding.

[82] Pinotti L, Dell'Orto V. Feed safety in the feed supply chain. BASE. 2011 Jan 1.

[83] Fields PG. The control of stored-product insects and mites with extreme temperatures. J Stored Prod Res 1992; 28(2): 89-118.
[http://dx.doi.org/10.1016/0022-474X(92)90018-L]

[84] Campabadal CA, Maier DE, Fahrenholz EB, Stark C, Jones C. Steam generation, control, and quality for feed manufacturing. Feed Pelleting Reference Guide 2014.

CHAPTER 7

Role of Biotechnology in Sustainable Mariculture

Syed Muzammil Munawar[1,*], **Dhandayuthabani Rajendiran**[1] and **Khaleel Basha Sabjan**[1]

[1] *Department of Chemistry & Biochemistry, C. Abdul Hakeem College (Autonomous) Melvisharam 632509, Ranipet District, Tamil Nadu, India*

Abstract: Marine biotechnology in pharmaceutical and food industries is a growing sector supported by expanding reach and financial resources worldwide. This study evaluates 620 publications and 29 projects, identifying the most promising technologies and business trends to advance materials development and improve performance. Omics, drug evaluation, and bioinformatics technologies are driving medical research and are considered essential tools to develop novel materials and candidate organisms for commercial applications. Techniques to optimize living conditions, harvesting, and purification methods, combined with recombinant technologies, are important to maximize commercial and industrial use, particularly in food and nutraceutical programs. Growth in this sector has been driven by increasing consumer demand for healthy ingredients that provide sustained health benefits, along with less stringent regulations compared to pharmaceutical products. Although this review has several limitations—particularly the author's subjectivity in classifying publications and an imperfect selection of projects analyzed—it offers insights into trends most relevant in Europe. It does not provide a fully quantitative assessment of the current literature in drug and food applications, but it highlights the development of basic research and its implementation during the period. As such, promoting evidence-based strategy creation is the primary recommendation for overcoming barriers that limit seafood commercialization.

Keywords: Aquaculture, Biotechnology, Food security, Public health, Sustainable development.

INTRODUCTION

Biotechnology is an important tool for the sustainable use of marine resources, as a series of cognitive and technological limitations hinders its full potential. Using a modern systematic approach that combines meta-analysis of 620 globally published items with 29 TRL (Technology Readiness Level) assessments from

* **Corresponding author Syed Muzammil Munawar:** Department of Chemistry & Biochemistry, C. Abdul Hakeem College (Autonomous) Melvisharam 632509, Ranipet District, Tamil Nadu, India;
E-mail: syed_bio2004@yahoo.co.in

G. Chelladurai, M. Paripooranaselvi & B. Makesh Kumar (Eds.)

EU-funded projects, this study focuses on evaluating the development opportunities of marine biotechnologies for pharmaceutical and nutraceutical applications, with an emphasis on technologies that address the major challenges in the commercialization of seafood products. The results show a trend of significant quality improvement, with rates more than doubling between 2010 and 2019. Biochemical and molecular characterization, represented in 150 studies, is a well-established technology. Emerging technologies in fundamental research include omics, drug evaluation, and bioinformatics, each of which has doubled in application over the last five years.

In contrast, technologies to optimize cultivation, harvesting, and extraction conditions are crucial for maximizing commercial potential, with 65% of the high TRL projects targeting immediate industrial exploitation, particularly in food and nutraceutical applications. This chapter provides preliminary guidance for studies aimed at overcoming restrictions that limit the marketing of organism-derived products. General trends in the development of aquaculture include (Subasinghe, 2007): (i) Continuing intensification of aquaculture production, (ii) Continuing diversification of species used, (iii) Continuing diversification of production structures and practices, (iv) Increasing market influence and consumer demand, (v) Enhancing regulations and governance in the sector, and (vi) Growing interest in higher control over production environments.

Aquaculture production is expected to increase, but the question remains whether it will grow sustainably and rapidly enough to meet demand while maintaining fish stocks. Biotechnology plays a key role in addressing this global challenge by supporting sustainable aquaculture development, economic and social progress, and environmental security worldwide. The application of biotechnology to the production of aquatic species has significant potential to improve aquaculture and meet the growing demand for aquatic food. As production of aquatic products increases, environmentally friendly strategies must be implemented to enhance productivity and improve product quality. At the same time, biotechnology contributes significantly to disease resistance and the rising value of aquatic species. By using disease-resistance genes, biotechnology reduces dependence on chemicals, particularly antibiotics, promoting sustainable aquaculture. It also provides effective tools to improve the health and safety of wild and farmed aquatic species, especially by increasing the proportion of fish suitable for commercial aquaculture.

Additionally, biotechnology allows the production of species in larger batches per unit area, reduces costs, supports biodiversity and critical ecosystems, and mitigates environmentally harmful aquaculture practices. However, the development of marine biotechnology in new areas faces challenges. Although several marine-derived products have been successfully commercialized through established technologies, the field remains marginal and underdeveloped in European and global markets. For the purposes of this chapter, the main challenges for advancing this industry and promoting new products are divided into three categories: discovering novel products, achieving sustainable production, and enhancing product performance and characteristics.

SUSTAINABLE PRODUCTION

When using marine organisms to produce valuable materials and biofuels, the vulnerability of marine ecosystems must be considered. Ready-to-use materials from marine organisms are often unsustainable for commercial applications due to the limited availability of some species. Many species are threatened, and overexploitation can destabilize fragile ecosystems, highlighting the need for alternative approaches. Traditional chemical and microbiological methods, including chemical combination or semi-synthesis, often have significant ecological impacts, particularly due to the use of toxic solvents and generation of large amounts of waste. Researchers and industrial operations in marine biotechnology must overcome these challenges by developing greener technologies [1].

Sustainable aquaculture strategies, including the cultivation of marine organisms such as corals and sponges, represent eco-friendly approaches, as these organisms provide important raw materials without further destabilizing marine ecosystems [2]. The environmental impact of producing bioactive compounds from marine organisms is often exacerbated by traditional production methods. For example, conventional microalgae production is energy-intensive, generates significant waste, and can account for up to 80% of total production costs. Expanding downstream capabilities through marine biotechnology provides a solution to these challenges [3].

Sustainable production can also leverage fishery waste to obtain high-value materials. Fish processing byproducts create significant waste—often exceeding 1,000 tons annually—which raises environmental and disposal concerns. The biochemical richness of this waste allows its conversion into valuable products through biotechnological methods, including protein hydrolysates with antioxidant properties or biomaterials such as collagen [4]. This approach reduces waste while providing a direct source of bioactive materials.

Moreover, the integration of green enzymatic processes in food production aims to reduce the use of toxic chemicals and synthetic reactions. Recently, nanoparticles have been developed for chemical, physical, and biomedical applications. Using algae or microorganisms as a production platform offers a sustainable alternative to traditional nanoparticle synthesis, which is environmentally harmful due to chemical usage [4]. Additional sustainable approaches, such as biocatalysis, benefit from enzymes isolated from marine organisms, which are adapted to extreme marine conditions. These enzymes often have unique stereochemical and catalytic properties compared to terrestrial counterparts. Notable examples include esterase from Yarrowia lipolytica, used to obtain racemic esters with strong antimicrobial activity, and proteases with specific substrate preferences suitable for peptide cyclization [5, 6].

GENETICALLY MODIFIED SPECIES – TRANSGENICS

The field of aquaculture has expanded into biotechnology by promoting genetic improvements in aquatic species. Transgenics, a key tool of modern biotechnology, involves creating new breeds by introducing specific genes to produce desirable traits, enhancing the genetic potential of species used in aquaculture. Transgenesis selectively transfers one or more genes to achieve advantageous changes, such as improved growth rate, size, feed-to-muscle conversion, and sexual maturity (El-Zaeems, 2004). Growth hormone genes from humans or animals have been introduced into species such as salmon, trout, and tilapia, resulting in faster growth compared to conventional breeding. When desired traits cannot be achieved through traditional breeding, genes can also be exchanged between species. Increased tolerance to environmental stress, extreme conditions, and low oxygen levels are important traits developed through transgenic approaches, including the introduction of Antifreeze Protein (AFP) genes to enhance survival in freezing environments (Hew *et al.*, 1992). Biotechnology also improves disease resistance, reducing variability between farmed and wild populations and supporting sustainable aquaculture (El-Zaeem and Aseem, 2004). Key techniques for gene transfer include microinjection into newly fertilized eggs and electroporation to deliver DNA into embryos or tissues.

Risks: Despite its potential, Genetically Modified Organisms (GMOs) face global acceptance challenges. Limited independent studies exist on human health risks, and horizontal gene transfer from GM plants to microorganisms has been documented (Snow *et al.*, 1999). Concerns also exist regarding allergenic proteins and ecological impacts, including escape into the environment, which could threaten wild populations and biodiversity (Tepfer, 2002).

HYBRIDIZATION

Hybridization involves crossing males of one species with females of another to combine desirable traits. In aquaculture, interspecific hybridization accelerates growth, integrates species-specific traits, and improves environmental resilience. It can also produce uniparental populations to achieve sexual dimorphism, as seen in tilapia. Stock restocking programs may use hybrids to reduce unwanted spawning, produce sterile fish, and increase tolerance to environmental stress. Hybridization is applied extensively in species such as striped bass in the United States, hybrid catfish in Thailand, and various salmonids, ensuring a steady supply to markets. Artificial insemination enables controlled reproduction, while sterile hybrids help prevent environmental impacts.

Risks: Escaped hybrids may pose ecological challenges, affecting genetic resources and biodiversity. Unexpected effects on hybrid offspring, such as reduced vigor or failure to achieve sterility, may also occur.

Hormonal Applications

Biotechnological tools, including Gonadotropin-Releasing Hormone (GnRH), are used to induce reproduction and early development in aquatic species (Alok *et al.*, 2000). Hormonal stimulation allows year-round gamete and juvenile production, supports domestication, and improves reproductive performance. Hormones can also induce infertility or triploidy to control reproduction and conserve endangered species.

Risks: Hormonal use poses potential threats to human and environmental health. Their effects on sexual dimorphism, fertility, and ecosystem toxicity are not fully understood.

CHROMOSOME SET MANIPULATION

Chromosome manipulation techniques, such as polyploidy induction, gynogenesis (maternal inheritance), and androgenesis (paternal inheritance), are widely used to control sex ratios, sterilize gonads, and produce homozygous lines in species such as tilapia, cyprinids, and salmonids. Triploids and tetraploid-diploid crosses ensure sterile offspring, facilitating sustainable aquaculture.

CRYOPRESERVATION

Cryopreservation preserves biological material at extremely low temperatures, typically -196°C in liquid nitrogen, halting physiological and biochemical activity. Cryoprotectants protect cells during freezing and thawing. This technology enables long-term storage of gametes, supports hybridization, mitigates low sperm counts, and allows year-round availability. Gene banks further safeguard the genetic resources and biodiversity of farmed and wild aquatic species.

AQUATIC SPECIES HEALTH AND VACCINES

Disease is a major obstacle to sustainable aquaculture, and biotechnology enhances health management. Tools such as gene probes and Polymerase Chain Reaction (PCR) improve pathogen detection. DNA vaccines, containing genes that produce antigens, provide targeted immunity and are more effective than traditional attenuated or killed vaccines. Non-specific immune stimulants, such as lipopolysaccharides, peptidoglycans, or glucans, also support health (Soltanian *et al.*, 2009). Advanced vaccines reduce chemical use, environmental contamination, and disease-related economic losses.

Risks: DNA vaccines may integrate into host genomes, amplify genes, form tumors, or affect non-target species. Genetic recombination and hybrid viral progeny are additional concerns, requiring careful monitoring.

FEED PRODUCTION

Nutrition is critical for maintaining healthy and productive aquaculture. As feed costs rise and fishmeal sources decline, biotechnology is applied to develop plant-based protein alternatives, optimize feed composition, and enhance digestibility. Excess phosphorus in feeds can cause eutrophication, highlighting the need for sustainable feeding strategies. Biotechnological advances, including enzyme supplementation, improve the utilization of plant-based diets and contribute to sustainable aquaculture practices.

Like other environmental challenges, biotechnology is being applied today to develop eco-friendly solutions for phosphorus pollution in aquaculture feeds.

DISCUSSION

Marine biotechnology is an emerging field focused on the sustainable use of marine bioresources by converting them into feedstocks or biofuels, often at high costs [7]. Many marine-derived products are currently being researched for their bioactive properties, and biotechnology is increasingly recognized as a sustainable approach to realizing these potentials. The growth of the sector is highly promising, with marine biotechnology projected to expand to $2.5 billion between 2020 and 2024 [8]. In Europe, the market reached USD 1,301.85 million, with pharmaceutical and food sectors accounting for more than 60% of the total value [9]. A major factor driving the marine biotechnology market is the vast biodiversity of marine ecosystems combined with the growing demand for alternative, health-promoting, and plant-based products.

To identify areas with the greatest potential for dominance in pharmaceutical and food products, researchers conducted a systematic review of literature from 2010–2020, focusing on projects funded between 2014 and 2020 that predominantly utilized marine biotechnology. A quantitative assessment of blue biotechnology literature over the past decade shows a significant increase in its use as a strategic tool to harness marine resources while combating ecosystem depletion. From a regulatory perspective, many countries have implemented frameworks to optimize marine ecosystem use, including initiatives aligned with the 1992 Convention on Biological Diversity and the United Nations Sustainable Development Goals. These regulations support market sustainability and stimulate both academic and private research in pharmaceutical and nutritional applications of marine biotechnology.

In China, although there is no strategy specifically dedicated to marine biotechnology, the sector has received government support through subsidies and tax incentives since 1996 as part of a broader economic growth strategy. Marine biotechnology is considered essential for "maximizing utilization of marine technology and production," according to Chinese government plans. The development of high-quality local institutions has further stimulated research in marine biotechnology, pharmaceutical, and food sectors. Notable centers include the Ocean University of China, with its laboratories in Marine Biotechnology, Marine Genetics, and Breeding, as well as the UNESCO Center of Marine Biotechnology. Additionally, the Chinese Academy of Sciences' Institute of Microbiology has contributed to numerous European projects, including the Pharmasea initiative, which explores novel MGATech bioactive molecules and enzyme sources from hypersaline environments [10, 11].

India presents a unique case among Asian countries. Since 1988, the authorities of the Department of Biotechnology have been implementing national and international projects focused on aquaculture and marine biotechnology. These initiatives are considered essential for utilizing biological resources and enhancing societal benefits. The program also includes the expansion of the National Maritime Center, which coordinates research and development efforts, though some aspects are still awaiting implementation.

Among European countries, Italy leads in the number of clinical trials and contributes significantly to the European strategy for blue growth. Since 2012, Italy has been a major contributor to the European Commission's initiatives promoting sustainable growth in areas such as aquaculture, renewable energy from oceans, coastal tourism, mineral resources, and marine biotechnology. This has led to the commercialization of marine biotechnology within the Italian Bioeconomy (BIT) and the emergence of Blue Italian Growth (BIG) technologies, emphasizing the sustainable development of marine bioresources. More than 30% of Italian research is conducted with support from the Zoological Station Anton Dohrn, one of Italy's leading marine science and biotechnology centers, with multidisciplinary expertise in ecology, physiology, genomics, transcriptomics, biochemistry, and molecular biology.

Despite its wide range of applications, marine biotechnology remains a developing field, often linked to experimental research rather than fully commercialized food products. A systematic literature review identified three main macrodomains or challenges, each related to the discovery of marine organisms with bioactive compounds to improve yield, efficiency, product properties, and sustainable production, particularly in waste management. About 65% of selected publications focus on fundamental research, searching for novel bioactive compounds, rather than on scaling production sustainably. Within twelve categorized strategies, the majority—biochemical and molecular characterization—focus on evaluating newly discovered organisms, enzymes, and molecules, ranking highest in innovation efforts. While foundational, this approach does not fully meet the costs or technological demands of innovative drug discovery and commercialization.

Other applied technologies focus on optimizing cultivation conditions and harvesting/extraction strategies for sustainability. Analysis of technology usage in selected studies shows a balance in favor of fundamental research, though it does not always align with current trends in marine biotechnology expansion. Consequently, there has been growing research on bioinformatics, omics technologies, and pharmacological evaluation, whose use doubled between 2015 and 2019 compared to the preceding five years. These technologies are primarily used to discover new marine organisms or compounds and are mostly implemented by universities and research institutes. Pharmacological evaluations involve molecular growth inhibition studies and antimicrobial assessments to quantify bioactivity, with some protective evaluations. The demand is for high-performing, safe, and human-relevant products, driving technology development. Omics and bioinformatics increasingly address the frontiers of marine research, including DNA, metabolic, and protein profiling, metagenomic libraries for uncultured samples, and *in silico* methods such as molecular modeling and docking. These strategies often integrate large datasets, particularly from metagenomics, supported by European-level funding for fundamental marine biotechnology projects.

Between 2014 and 2020, the European Commission allocated €149,433,491 for blue growth activities, including food and pharmaceuticals, under Horizon 2020 programs related to food security, agriculture, sustainable forestry, marine services, maritime and inland waters, and bioeconomy research. The analysis highlights that many promising lab-scale technologies are not yet commercially viable. High-TRL projects, mainly funded by the SME Instrument, focus on assessing business potential, optimizing cultivation and extraction strategies, and managing raw materials, representing 65% of projects that are too advanced to fully integrate sustainability and productivity. Emerging areas include new biorefineries and sustainable aquaculture systems for surplus food, large-scale microalgae cultivation using modern bioreactors, and production strategies aimed at improving animal welfare.

Promising technologies under ongoing research include rDNA methods to enhance fish meat quantity and quality, and the development of seaweed-based products. These strategies are primarily applied to current projects, emphasizing both commercial potential and innovation. Recombinant technology combined with optimization strategies represents a highly promising approach, capable of overcoming many commercialization challenges while ensuring performance and longevity. Products with market appeal, such as beneficial ingredients and nutraceuticals, are being developed to meet consumer demand for sustainable food production and improved nutrition, complementing terrestrial crops. Strategies for microalgae cultivation maximize nutraceutical value while minimizing ecological impact, integrating seamlessly into the environment, and optimizing resource use.

The Blue Iodine II project exemplifies a sustainable business model, utilizing algae biomass in biorefineries with partial energy input and recycling of fishery by-products as a closed-loop process. By promoting sustainability throughout the supply chain, including packaging of nutraceuticals, the marine biotechnology sector is expanding rapidly, supporting eco-friendly food production, enhancing animal health, and enabling high-value production. The development of sustainable food and feed products in marine biotechnology highlights key opportunities and identifies the most promising industries for future investment.

Weighing qualitative and quantitative assessments of market characteristics, the present analysis provides not only an effective description of the marine biotechnology landscape but also offers practical frameworks that serve as starting points for overcoming barriers limiting the commercialization of products derived from marine organisms. For research outputs to meet market demands, consistent communication between research laboratories and companies is essential. This vision unifies laboratories and industry, bridging gaps in environmental commercialization collaborations. Such collaboration allows researchers to access market data, identify the most promising technologies and products, pursue further market opportunities, and connect with companies aligned with emerging market trends.

Highlighting specific commodities as drivers of the future marine biotechnology market provides the industry with a powerful tool to identify the most promising sectors, particularly in the foodservice space. This approach enables the evaluation of trading companies and commercial models that respond directly to market needs. These models support the development of sustainable food resources, address consumer willingness to pay a premium for predominantly marine-derived products, and integrate fully with environmental and resource optimization strategies.

Nationally, selected manufacturers and policymakers in marine biotechnology are enacting regulations to standardize research, guide fund allocation, and implement rules that support the sustainability of marine resources. In particular, the Food Industry is leveraging these initiatives to pursue new food products that meet changing consumer demands, thereby promoting market adoption of marine biotechnology solutions. The assessment presented in this chapter further supports evidence-based policymaking by defining new positional datasets, establishing planning rules, and proposing indicators across key financing sectors of the blue economy [12 - 14].

REFERENCES

[1] Leary D, Vierros M, Hamon G, Arico S, Monagle C. Marine genetic resources: A review of scientific and commercial interest. Mar Policy 2009; 33(2): 183-94. [http://dx.doi.org/10.1016/j.marpol.2008.05.010]

[2] Mora C, Tittensor DP, Adl S, Simpson AGB, Worm B. How many species are there on Earth and in the ocean? PLoS Biol 2011; 9(8): e1001127. [http://dx.doi.org/10.1371/journal.pbio.1001127] [PMID: 21886479]

[3] OECD. Marine Biotechnology Enabling Solutions for Ocean Productivity and Sustainability. Paris, France: OECD Publishing 2013.

[4] Wijffels RH, Barbosa MJ. An outlook on microalgal biofuels. Science 2010; 329(5993): 796-9. [http://dx.doi.org/10.1126/science.1189003] [PMID: 20705853]

[5] Suleria H, Osborne S, Masci P, Gobe G. Marine-based nutraceuticals: An innovative trend in the food and supplement industries. Mar Drugs 2015; 13(10): 6336-51. [http://dx.doi.org/10.3390/md13106336] [PMID: 26473889]

[6] Subasinghe RP. Aquaculture: Status and Prospects. Role of Aquaculture in Sustainable Development. Rome, Italy: FAO Department of Fisheries and Aquaculture 2007.

[7] El-zaeems SY. Alteration of the productive performance characteristics of Orechromis niloticus and Tilapia Zillii under the effect of foreign DNA injection. Egypt J Aquat Boil Fish 2004; 8(1): 261-78.

[8] El-Zaeem, S.Y., Aseem, S.S. 2004. Application of biotechnology in fish breeding: 1 - production of highly immune genetically modified Nile, tilapia Orechromis niloticus with accelerated growth by direct injection of Shark.

[9] Hew CL, Davies PL, Fletcher G. Antifreeze protein gene transfer in Atlantic salmon. Mol Mar Biol

Biotechnol 1992; 1(4-5): 309-17.
[PMID: 1308821]

[10] Snow AA, Andersen BENTE, Jørgensen RB. Costs of transgenic herbicide resistance introgressed from *Brassica napus* into weedy *B. rapa*. Mol Ecol 1999; 8(4): 605-15.
[http://dx.doi.org/10.1046/j.1365-294x.1999.00596.x]

[11] Tepfer M. Risk assessment of virus-resistant transgenic plants. Annu Rev Phytopathol 2002; 40(1): 467-91.
[http://dx.doi.org/10.1146/annurev.phyto.40.120301.093728] [PMID: 12147768]

[12] Alok D, Talwar GP, Garg LC. *In vivo* activity of salmon gonadotropin releasing hormone (GnRH), its agonists with structural modifications at positions 6 and 9, mammalian GnRH agonists and native cGnRH-II on the spawning of an Indian catfish. Aquacult Int 2000; 7(6): 383-92.
[http://dx.doi.org/10.1023/A:1009262230465]

[13] Bioactive compounds from Marine sources. Available from: http;//docplayer.net/46861255-Biotechnology-and-aquaculture-in-sustainable-development.html

[14] Marine Biotechnology: Challenges and Development Market Trends for the Enhancement of Biotic Resources in Industrial Pharmaceutical and Food Applications. A Statistical Analysis of Scientific Literature and Business Models. Available from: http//pubmed.ncbi.nlm.nih.gov/33430360/

CHAPTER 8

Planning and Regulation of Sustainable Mariculture

P. J. Joslin[1,*], **M. I. Delighta Mano Joyce**[2] and **M. Paripooranaselvi**[1]

[1] *Department of Zoology, St. Mary's College (Autonomous), Thoothukudi-628001, Tamil Nadu, India*

[2] *Department of Zoology, Sadakathullah Appa College (Autonomous), Tirunelveli, Tamil Nadu, India*

Abstract: India is viewed as a promising landscape for expanding mariculture and addressing the numerous challenges faced by the food industry. Despite the many benefits of India's green, blue, and silver revolutions, intensive aquaculture techniques have raised environmental concerns in aquatic environments, causing significant concern. Improved farm management practices, integrated farming, selective chemical usage, proper disease management, the development of low-cost and novel technologies, technology transfer for feed management, and other methods could help mariculture achieve long-term sustainability.

Keywords: Mariculture, Planning, Regulation, Sustainability.

INTRODUCTION

Aquaculture is a specialized form of farming that involves the controlled or semi-controlled cultivation of marine organisms for food and other purposes in coastal and offshore waters. Potentially cultivated species in India include fish, crustaceans, molluscs, algae, and ornamental fish. The development of aquaculture in India is gaining significant momentum, thanks to innovative technologies. However, environmentally friendly approaches must be adopted when striving to improve and expand the field. Aquaculture should be practiced as part of sustainable development. Sustainable polyculture is the technique of raising two or more compatible aquatic species in a single pond or lake, with the aim of increasing production by utilizing organisms with different feeding habits or spatial distributions [1]. Traditionally, mariculture has taken place at the interface of land and sea, in tidal flats, estuaries, and sheltered bays. Although

* **Corresponding author P. J. Joslin:** Department of Zoology, St. Mary's College (Autonomous), Thoothukudi 628001, Tamil Nadu, India; E-mail: jos.mathias67@gmail.com

G. Chelladurai, M. Paripooranaselvi & B. Makesh Kumar (Eds.)

calm waters and easy access make nearshore mariculture attractive, environmental impacts and conflicts with other uses have increased in these increasingly populated coastal areas. Advances in technology and farming methods have allowed farms to be established far from the coast and in harsher ocean conditions, opening up new potential aquaculture areas [2, 3]. Offshore aquaculture promises to increase seafood supply and serve as a new source of economic development.

ENVIRONMENTAL MONITORING PLANS

This plan outlines the mitigation, management, and monitoring methods to be applied during project development and operation to reduce or eliminate negative environmental impacts to a manageable level. By allocating the necessary funds, it ensures that environmental factors are safeguarded throughout various activities in the production process. It also provides for periodic monitoring of vital parameters, employee health, and all required safety and environmental measures. The committee ensures that management communicates relevant environmental factors and impacts to the implementation teams and members of the Environmental Health and Safety (EHS) unit in a clear and organized manner, without omitting any important information. This process helps clarify and streamline the EHS unit's role.

INTEGRATED MULTI-TROPHIC AQUACULTURE (IMTA)

Integrated Multitrophic Aquaculture (IMTA) involves farming species at different trophic levels in close proximity, with complementary ecosystem functions. In this system, the food, waste, nutrients, and by-products of one species are utilized and converted into fertilizer, feed, and energy for other species. It also takes advantage of synergistic interactions between species while providing biological attenuation. Algae and aquatic plants, which remove dissolved inorganic nutrients, along with shellfish and other invertebrates, which remove organic particulate nutrients for growth, are examples of extractive species cultivated alongside fed species such as shrimp and fish. The aim is to design hydroponic systems that extend natural organic management, support economic stability by increasing production, reducing costs, diversifying products, minimizing risks, and creating new job opportunities, as well as promoting social acceptance.

SUSTAINABLE FEED MANAGEMENT

The goal of food management in aquaponic systems is to provide fish and plants with appropriate nutrition while controlling input costs and minimizing pollutants. Effective food management involves determining when and how often to feed, how much to provide, and the most efficient feeding methods. While nutrition is a key factor in feed planning, optimal feed management also has a substantial impact on production costs and the overall health of the system. Feed costs and environmental impacts are therefore critical considerations in dietary management.

SUSTAINABLE USE OF CHEMICALS

The use of chemotherapy, synthetic compounds, and various drugs has been introduced in Indian aquaculture. Numerous products are employed for diverse purposes, including soil and water treatment chemicals, disinfectants, piscicides, herbicides, natural and synthetic fertilizers, food additives, therapeutic drugs, and sedatives. Methods involving synthetic materials carry potential hazards and impacts associated with the use of these compounds. Various disease management strategies—such as habitat planning, vegetation density control, and effective treatment frameworks—are reviewed, along with public guidelines on the safe use of these compounds. Research on aquaculture systems and chemical applications in India is ongoing. Suggestions for improved and responsible use of synthetic compounds in Indian aquaculture are being sought by breeders, government agencies, aquaculture organizations, the chemical industry, local and international bodies, and recognized research institutions.

PRECISION FISH FARMING

This type of fish farming utilizes various sensors to collect data on the farm's environmental conditions, enabling informed decisions that enhance fish welfare, growth, and financial performance while mitigating climate-related risks. These technological advances have also influenced horticulture, where sensors and related innovations enhance knowledge of plant health and receive government support for agricultural development [4]. Traditionally, many of these processes have relied heavily on human intervention and farmers' expertise to make the right decisions and take appropriate action. However, as farm sizes grow and operations expand, automation becomes essential for maintaining economic viability. Wireless communication systems are particularly important, as fish farms often include extensive chains, ropes, anchors, and other submerged equipment. Additionally, the industry's fragmented structure—characterized by numerous small-scale aquaculture companies and sensor suppliers—creates challenges in integrating diverse and sometimes proprietary data sets into a unified edge-fog-cloud ecosystem.

DISEASE MANAGEMENT

Successful fish health management begins with disease prevention rather than treatment. Effective prevention requires proper water quality management, adequate nutrition, and strict adherence to hygiene practices. Without these measures, opportunistic outbreaks cannot be avoided. Fish are continually exposed to potential pathogens, including bacteria, fungi, and parasites. Even advanced disinfection technologies, such as ultraviolet treatment or ozonation, cannot eliminate all of them from the environment. Suboptimal water quality, poor nutrition, or stress-induced immune suppression can enable these pathogens to cause disease.

Bacterial Diseases

Bacterial diseases often manifest as streaks, red spots, or swelling in the abdomen or eyes. When infections occur, antibiotics such as penicillin, amoxicillin, or erythromycin may be administered for treatment.

Fungal Diseases

Common fungal infections typically appear as gray or white, fluffy patches.

Parasitic Disease

The most common parasitic disease, known as "Ich," can be effectively treated with an appropriate dose of copper or malachite green. Most treatments use copper as the active ingredient, and many commercial water treatments, such as Aquari-Sol, also contain copper. When administering antibiotics or copper-based treatments, all carbon should be removed from the filtration system to prevent it from adsorbing the medication.

Environmental Factors Often Cause Physical Discomfort in Fish

Poor water quality can lead to symptoms such as gasping for air, loss of appetite, or jumping out of the aquarium. Stress or difficulty adapting to the environment may also result in aggressive behaviors, including biting and fin nipping.

AN AWAKENING

The Blue Revolution Endowment Fund, also known as the Neel Kranti Mission, aims to improve the country's economic prosperity by enhancing fishing and fish farming, while contributing to food security and nutrition. Its mission is to utilize water resources to develop fisheries sustainably, with careful consideration of biosecurity and environmental concerns.

DEVELOPMENT OF SUSTAINABLE METHODS

Site Selection

One of the main challenges to the sustainable development of aquaculture is the allocation of water, land, and other resources for purposes other than aquaculture, such as fishing, agriculture, and tourism. Determining suitable locations for farm sites, as well as proper zoning for aquaculture, must be considered in marine spatial planning [5]. The negative environmental impacts of aquaculture can be mitigated by selecting ecologically appropriate sites away from sensitive ecosystems and ensuring proper regulatory oversight. Therefore, aquaculture operators must maintain a clean environment to support the growth of healthy fish. Geographic Information Systems (GIS), remote sensing, and modeling are commonly employed to select aquaculture locations, as they provide valuable preliminary information for on-site assessments.

Recirculating Aquaculture Systems (RAS) allow fish to be cultured under controlled conditions, reduce the risk of parasitic diseases, and adapt to changes in environmental circumstances, ensuring effective waste management. To achieve optimal results, site selection must be carefully planned and aligned with the objectives of marine spatial planning, integrating the economic, social, and environmental goals of each area.

INTEGRATED MULTITROPHIC AQUACULTURE (IMTA)

Active ingredients in aquaculture contain numerous essential nutrients and unsaturated fats that should be recycled rather than discarded within a sustainable aquaculture system [6]. To address this, the Integrated Multi-Trophic Aquaculture (IMTA) concept was developed, incorporating an innovative food web structure that integrates the cultivation of species such as fish and shrimp with organisms like molluscs and kelp, which can capture particles and nutrients from the environment. Despite growing consumer demand for environmentally sustainable practices, IMTA systems can achieve higher yields than single-species systems [7]. This approach applies the principle of combining species from different trophic levels to create a more balanced and ecologically efficient farming system.

SUSTAINABLE FOOD MANAGEMENT

Control of aquaculture feed is a critical issue affecting the sustainability of marine fish farming. New methods, including the use of underwater cameras, have been developed to reduce excess feed and minimize feed loss. It is essential to monitor feeding activity and adjust the amount of food according to the needs of fish in intensive farming systems, where feeding is carried out according to a predetermined schedule. Underwater camera technology can detect unconsumed feed and automatically stop feeding, providing an effective tool for managing and monitoring feed.

CULTIVATION OF ALGAL SPECIES

When using water-degradable nitrogen to promote the growth of algae or microalgae, fertilizers or additional growth media are not required. Different species can then be cultivated as a food source, enhancing their economic value. For example, the macroalga *Ulva lactuca* has limited direct market value but serves as excellent feed for abalone and other marine species, making green algae cultivation a more viable alternative to traditional fishmeal.

Marine algae farming provides a range of direct and indirect benefits, including creating jobs for coastal communities while posing minimal risk to the ecosystem, supplying kelp in sufficient quantities for various industrial uses, providing a continuous source of raw materials for kelp-based businesses, delivering consistent-quality biomass for industry, and reducing pressure on conventional resources. Additionally, bioremediation digesters can treat sludge and wastewater from aquaculture operations, and their biofilms, which contain biomass, can be repurposed as a food source.

BIOFLOC TECHNOLOGY

Biofloc systems are commonly used in freshwater aquaculture, particularly for herbivorous species, although recent applications have expanded to include Pacific mullet and shrimp. In this approach, the addition of carbohydrates—such as sugar, starch, or cellulose—promotes microbial growth in the water. Heterotrophic bacteria utilize these carbohydrates as substrates to produce proteins essential for growth, while simultaneously assimilating nitrogen from the surrounding water. As a result, nitrogen compounds are fully converted into bacterial biomass, forming biofloc that shrimp and fish can consume.

PRECISION FISH FARMING (PFF)

Precision Fish Farming (PFF) is an economic and management approach to aquaculture, adapted from the concept of Precision Livestock Farming (PLF) [8]. Both PLF and PFF utilize devices, such as sensors, monitoring systems, and software, to enhance animal welfare and productivity. In PFF, tools such as smart sediment traps can provide detailed information on sediment fluxes, nutrient cycles, and feed efficiency through site-based measurements and the detection of sedimentation patterns.

Disease management is a critical component of PFF. The use of antibiotics has raised concerns regarding the emergence and selection of antibiotic-resistant microorganisms. Proper training on antibiotic administration and awareness of potential negative effects are essential. Proactive monitoring using appropriate diagnostic tools is often the most effective strategy to prevent disease outbreaks. Stress reduction should also be prioritized through improved health management practices, and vaccination remains a highly effective method for preventing various infectious diseases. Vaccines can be administered orally, by injection, or through immersion or spraying. Pesticides and piscicides are sometimes used to eliminate harmful species from the ecosystem; however, residues can persist in water and sediment for weeks, often affecting non-target organisms. Reduced stocking densities, adequate spacing between farms, preventive measures, and good management practices can help minimize the use of chemical pesticides. Since many diseases are absent in cultured species, prevention should be emphasized as the primary strategy to control disease transmission. Enhanced surveillance programs for both known and emerging diseases, along with the use of biomolecular diagnostic technologies, should be promoted. Additionally, strict regulations on quarantine and animal movement are essential to prevent the spread of infectious diseases [9].

SPATIAL MANAGEMENT TOOL

Coastal and Marine Spatial Planning (MSP) is a science-based strategy designed to address specific challenges in ocean management while setting economic and environmental goals. Using Geographic Information Systems (GIS), ICAR-CMFRI has developed an innovative decision-support model to identify potential mariculture sites [10]. To create a comprehensive model that identifies optimal locations for mariculture operations, multiple parameters related to the coastal zone are assessed, including topography, physical and oceanographic characteristics, biological factors, and social infrastructure. The methodology is based on criteria that evaluate the structural and biological feasibility of deep-sea fish farming systems along entire coastlines. Currently, this model is being implemented for the coastal waters of Gujarat. However, it can also be adapted for other Indian coastal states, contributing to sustainable, ecosystem-based marine environmental management and the implementation of MSP in India.

INFLUENCE OF EXTERNAL MARKETS

Aquaculture in India is strongly oriented toward the international market, with exports being a key priority. In many cases, short-term economic benefits take precedence over long-term social, ecological, and environmental considerations. Therefore, strict regulations on licensing, certification, and promotion of large-scale projects targeting foreign markets are essential. Another potential drawback of international market influence is the neglect of domestic markets [11]. Emphasis should be placed on promoting and developing local markets for farmed species. With proper support, India can become a hub for innovative, sustainable, small-scale mariculture operations, providing positive impacts on livelihoods and food security at the local level.

CONCLUSION

Although Indian research institutes have conducted numerous trials and experiments on aquaculture technologies to improve the livelihoods of disadvantaged communities, they still face limitations in capacity development, strategic planning, effective stakeholder engagement, communication, and technology transfer. For marine aquaculture to be sustainable and impactful, technology development must focus on cost-effective solutions that are environmentally friendly, technologically advanced, and require minimal technical expertise during the operational phase.

REFERENCES

[1] Zimmermann S, Nair CM, New MB. Grow-Out Systems-Polyculture and Integrated Culture. Freshwater Prawns: Biology and Farming. Wiley- Blackwell 2009; pp. 195-217. [http://dx.doi.org/10.1002/9781444314649.ch11]

[2] Naylor R, Goldburg R. Future seascapes, fishing, and fish farming. Frontiers in Ecology 2012; 3: 21-8.

[3] Shainee M, Haskins C, Ellingsen H, Leira BJ. Designing offshore fish cages using systems engineering principles. Syst Eng 2012; 15(4): 396-406. [http://dx.doi.org/10.1002/sys.21200]

[4] Føre M, Frank K, Norton T, *et al.* Precision fish farming: A new framework to improve production in aquaculture. Biosyst Eng 2018; 173(173): 176-93. [http://dx.doi.org/10.1016/j.biosystemseng.2017.10.014]

[5] Chen Y, Del Genio AD, Chen J. The tropical atmospheric El Niño signal in satellite precipitation data and a global climate model. J Clim 2007; 20(14): 3580-601. [http://dx.doi.org/10.1175/JCLI4208.1]

[6] Bischoff JF, Rehner SA, Humber RA. A multilocus phylogeny of the *Metarhizium anisopliae* lineage. Mycologia 2009; 101(4): 512-30. [http://dx.doi.org/10.3852/07-202] [PMID: 19623931]

[7] Neori A, Chopin T, Troell M, *et al.* Integrated aquaculture: rationale, evolution and state of the art emphasizing seaweed biofiltration in modern mariculture. Aquaculture 2004; 231(1-4): 361-91. [http://dx.doi.org/10.1016/j.aquaculture.2003.11.015]

[8] Norton T, Berckmans D. Engineering advances in Precision Livestock Farming Biosystems. 2018; 173: 1-3.

[9] Available from: https://agriallis.com./wp-content/uploads/2021/06/FISH-DISEASE-AND-HE-LTH-MANAGEMENT.Pdf

[10] Divu D. Vase, Vinay Kumar and Kumar, Rajan and Nakhawa, A D and Rahangdale, Shikha and Jayasankar, J and Koya, Mohammed and and Rohit, Prathibha. Fishery and estimation of potential yield of Bombay duck Harpadonnehereus (Hamilton, 1822) along Gujarat coast, India. Indian J Fish

2020; 67(1): 1-7.

[11] Offshore Aquaculture: Spatial planning principles for sustainable development. Available from: https://pubmed.ncbi.nlm.nig.gov//2816067/

SUBJECT INDEX

A

B

C

D

G. Chelladurai, M. Paripooranaselvi & B. Makesh Kumar (Eds.)

E

F

G

H

I

L

M

www.ingramcontent.com/pod-product-compliance
Lightning Source LLC
LaVergne TN
LVHW070953210926
848482LV00002B/199
9798898816599